Lecture Notes in Applied and Computational Mechanics

Volume 64

Series Editors

Prof. Dr.-Ing. Friedrich Pfeiffer
Prof. Dr.-Ing. Peter Wriggers

Lecture Notes in Applied and Computational Mechanics

Edited by F. Pfeiffer and P. Wriggers

Further volumes of this series found on our homepage: springer.com

Vol. 64 Zohdi, T.I.
Electromagnetic Properties of Multiphase Dielectrics
180 p. 2012 [978-3-642-28426-7]

Vol. 63 Langer, U., Schanz, M., Steinbach, O., Wendland, W. L. (Eds.)
Fast Boundary Element Methods in Engineering and Industrial Applications
272 p. 2012 [978-3-642-25669-1]

Vol. 61 Frémond, M., Maceri, F. (Ed.)
Mechanics, Models and Methods in Civil Engineering
498 p. 2012 [978-3-642-24637-1]

Vol. 59 Markert, B., (Ed.)
Advances in Extended and Multifield Theories for Continua
219 p. 2011 [978-3-642-22737-0]

Vol. 58 Zavarise, G., Wriggers, P. (Eds.)
Trends in Computational Contact Mechanics
354 p. 2011 [978-3-642-22166-8]

Vol. 57 Stephan, E., Wriggers, P.
Modelling, Simulation and Software Concepts for Scientific-Technological Problems
251 p. 2011 [978-3-642-20489-0]

Vol. 54: Sanchez-Palencia, E., Millet, O., Béchet, F.
Singular Problems in Shell Theory
265 p. 2010 [978-3-642-13814-0]

Vol. 53: Litewka, P.
Finite Element Analysis of Beam-to-Beam Contact
159 p. 2010 [978-3-642-12939-1]

Vol. 52: Pilipchuk, V.N.
Nonlinear Dynamics: Between Linear and Impact Limits
364 p. 2010 [978-3-642-12798-4]

Vol. 51: Besdo, D., Heimann, B., Klüppel, M., Kröger, M., Wriggers, P., Nackenhorst, U.
Elastomere Friction
249 p. 2010 [978-3-642-10656-9]

Vol. 50: Ganghoffer, J.-F., Pastrone, F. (Eds.)
Mechanics of Microstructured Solids 2
102 p. 2010 [978-3-642-05170-8]

Vol. 49: Hazra, S.B.
Large-Scale PDE-Constrained Optimization in Applications
224 p. 2010 [978-3-642-01501-4]

Vol. 48: Su, Z.; Ye, L.
Identification of Damage Using Lamb Waves
346 p. 2009 [978-1-84882-783-7]

Vol. 47: Studer, C.
Numerics of Unilateral Contacts and Friction
191 p. 2009 [978-3-642-01099-6]

Vol. 46: Ganghoffer, J.-F., Pastrone, F. (Eds.)
Mechanics of Microstructured Solids
136 p. 2009 [978-3-642-00910-5]

Vol. 45: Shevchuk, I.V.
Convective Heat and Mass Transfer in Rotating Disk Systems
300 p. 2009 [978-3-642-00717-0]

Vol. 44: Ibrahim R.A., Babitsky, V.I., Okuma, M. (Eds.)
Vibro-Impact Dynamics of Ocean Systems and Related Problems
280 p. 2009 [978-3-642-00628-9]

Vol.43: Ibrahim, R.A.
Vibro-Impact Dynamics
312 p. 2009 [978-3-642-00274-8]

Vol. 42: Hashiguchi, K.
Elastoplasticity Theory
432 p. 2009 [978-3-642-00272-4]

Vol. 41: Browand, F., Ross, J., McCallen, R. (Eds.)
Aerodynamics of Heavy Vehicles II: Trucks, Buses, and Trains
486 p. 2009 [978-3-540-85069-4]

Vol. 40: Pfeiffer, F.
Mechanical System Dynamics
578 p. 2008 [978-3-540-79435-6]

Vol. 39: Lucchesi, M., Padovani, C., Pasquinelli, G., Zani, N.
Masonry Constructions: Mechanical Models and Numerical Applications
176 p. 2008 [978-3-540-79110-2]

Vol. 38: Marynowski, K.
Dynamics of the Axially Moving Orthotropic Web
140 p. 2008 [978-3-540-78988-8]

Electromagnetic Properties of Multiphase Dielectrics

A Primer on Modeling, Theory and Computation

Tarek I. Zohdi

Springer

Author
Tarek I. Zohdi
University of California
Berkeley, California
USA

ISSN: 1613-7736 e-ISSN: 1860-0816
ISBN: 978-3-642-28426-7 e-ISBN: 978-3-642-28427-4
DOI 10.1007/ 978-3-642-28427-4
Springer Heidelberg New York Dordrecht London

Library of Congress Control Number: 2012932278

Printed on acid-free paper

Springer is part of Springer Science+Business Media (www.springer.com)

Dedicated to my father, Prof. Magd E. Zohdi, who has given me guidance throughout my life. All profits generated by this book will be donated to the United Nations Children's Fund (UNICEF).

 I am certain that, despite painstaking efforts, there remain errors of one sort or another in this monograph. Therefore, I would be grateful if readers who find such flaws would contact me at *zohdi@me.berkeley.edu*.

Preface

Recently, several applications, primarily driven by microtechnology, have emerged where the use of materials with tailored electromagnetic (dielectric) properties are necessary for a successful overall design. The "tailored" aggregate properties are achieved by combining an easily moldable base matrix with particles having dielectric properties that are chosen to deliver (desired) effective properties. In many cases, the analysis of such materials requires the simulation of the macroscopic and microscopic electromagnetic response, as well as its resulting coupled thermal response, which can be important to determine possible failure in "hot spots." This can necessitate a stress analysis, in conjunction with models of chemical processes, to assess the damage occurring in the material. A central objective of this work is to provide basic models and numerical solution strategies to analyze the coupled response of such materials by direct simulation using standard laptop/desktop equipment. Accordingly, this monograph covers:

1. the foundations of Maxwell's equations,
2. basic homogenization theory,
3. coupled systems (electromagnetic, thermal, mechanical and chemical),
4. numerical methods and
5. an introduction to select biological problems.

The text can be viewed as a research monograph suitable for use in an upper-division undergraduate or first year graduate course geared towards students in the physical sciences, engineering and applied mathematics that have an interest in the analysis of new materials.

Berkeley, USA, January 2012 *Tarek I. Zohdi*

Contents

List of figures

1

Multiphase Continua, an Introduction

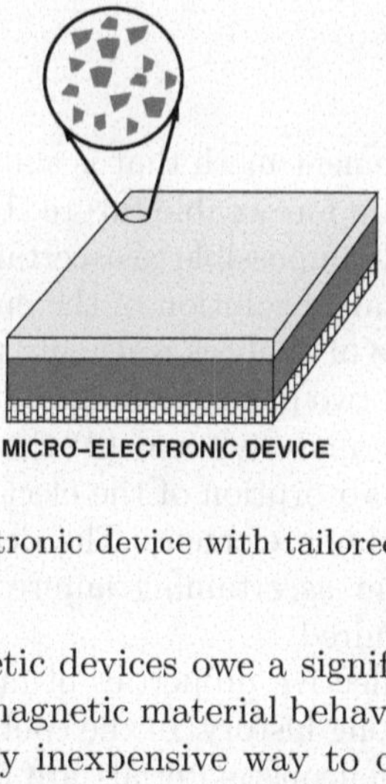

Fig. 1.1. A microelectronic device with tailored layers (Zohdi [137]).

Most modern electromagnetic devices owe a significant amount of their success to the tailored electromagnetic material behavior of the components that comprise them. A relatively inexpensive way to obtain macroscopically desired responses is to enhance an easy-to-form matrix material's properties by introducing microscale particles possessing different electromagnetic, thermal and mechanical properties. (Figure 1.1). The particles are chosen to produce an overall desired electromagnetic effect. The aggregate response of the material is an outcome of the interaction between the smaller-scale (microstructure) constituents that comprise the "effective" material. In the construction of such materials, the basic philosophy is to select matrix/particle material combinations in order to produce desired aggregate responses. For example, in electromagnetic engineering applications, the classical choice is to add a particulate phase with suitable dielectric constants in order to modify the overall properties of an easily moldable base matrix material.

Unfortunately, an attempt to directly simulate a device containing billions or even trillions of particles, incorporating all of the microscale details, requires an extremely fine spatial discretization mesh, for example, that of a

T.I. Zohdi: Electromagnetic Properties of Multiphase Dielectrics, LNACM 64, pp. 1–4.
springerlink.com

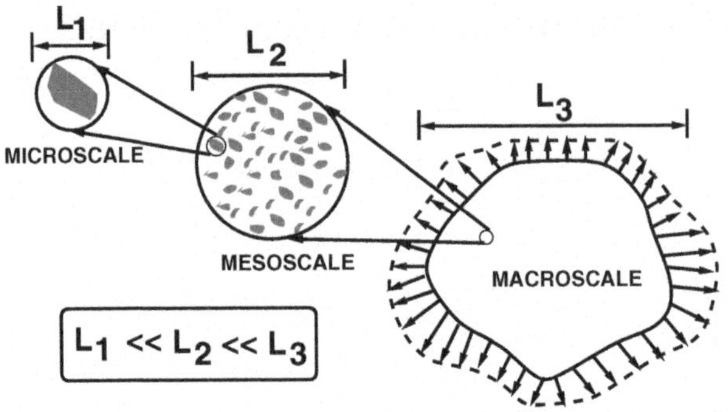

Fig. 1.2. A representative sample of a material with heterogeneous microstructure (Zohdi [137]).

finite difference or finite element mesh that is simply out of reach of virtually all computing devices in the foreseeable future. Furthermore, the exact subsurface geometry is typically impossible to ascertain throughout the structure. Essentially, a complete detailed solution of the entire device is nearly impossible. Fortunately, this type of analysis is usually unnecessary since, typically, an engineer wishes to know two primary things: (1) the effective or macroscale properties of the aggregate material, and (2) the resulting thermal response of the material due to the absorption of the electromagnetic energy in a statistically representative volume element. The determination of the thermal response is often critical in ascertaining failure of a device, thus, a stress analysis is also usually required.

The determination of effective properties of materials with heterogeneous microstructure has had a long history. In the context of electromagnetics, the properties of microheterogeneous materials are characterized by a spatially variable permittivity $\boldsymbol{\epsilon}$. Typically, in order to characterize the (homogenized) effective macroscopic response of such materials, a relation between averages

$$\langle \boldsymbol{D} \rangle_\Omega = \boldsymbol{\epsilon}^* \cdot \langle \boldsymbol{E} \rangle_\Omega, \tag{1.1}$$

is sought, where Ω is the domain, and

$$\langle \cdot \rangle_\Omega \stackrel{\text{def}}{=} \frac{1}{|\Omega|} \int_\Omega \cdot \, d\Omega \tag{1.2}$$

is the volume averaging operator and $\boldsymbol{D}$ and $\boldsymbol{E}$ are, respectively, the electric flux and electric field within a statistically representative volume element (RVE) of volume $|\Omega|$. The quantity, $\boldsymbol{\epsilon}^*$, is known as the effective permittivity property, and is the permittivity tensor used in usual macroscale analyses. Similarly, one can describe other effective quantities such as

$$\langle \boldsymbol{B} \rangle_{\Omega} = \boldsymbol{\mu}^{*} \cdot \langle \boldsymbol{H} \rangle_{\Omega}, \tag{1.3}$$

where $\boldsymbol{B}$ and $\boldsymbol{H}$ are the magnetic flux and magnetic field, and $\boldsymbol{\mu}^{*}$ is the effective magnetic permeability. For a sample to be statistically representative, it must usually contain a reasonably large number of particles (Figure 1.2) and, therefore, the computations over the RVE are still not trivial, but are of reduced computational effort in comparison with a direct attack on the entire macrostructural problem, if it were even possible.

Historically, because of the extreme difficulties in directly computing effective properties, a variety of approximate techniques has been developed to estimate the overall macroscopic properties of materials consisting of a matrix containing distributions of particles, dating back to at least Maxwell [68] (1867), [69] (1873) and Lord Rayleigh [90] (1892). One can generate rigorous bounds on the effective electromagnetic response. For example,

$$\langle \boldsymbol{\epsilon}^{-1} \rangle_{\Omega}^{-1} \leq \boldsymbol{\epsilon}^{*} \leq \langle \boldsymbol{\epsilon} \rangle_{\Omega}, \tag{1.4}$$

where the upper bound is generated by assuming that the electric field is uniform throughout the medium and the lower bound is generated by assuming that the electric field flux is uniform throughout the medium. For the magnetic properties, one has

$$\langle \boldsymbol{\mu}^{-1} \rangle_{\Omega}^{-1} \leq \boldsymbol{\mu}^{*} \leq \langle \boldsymbol{\mu} \rangle_{\Omega}, \tag{1.5}$$

where the upper bound is generated by assuming that the magnetic field is uniform throughout the medium and the lower bound is generated by assuming that the magnetic field flux is uniform throughout the medium. In the electromagnetics literature, the bounds in Equations 1.4 and 1.5 are usually referred to as the Wiener bounds [109]. These inequalities mean that the eigenvalues of the tensors $\boldsymbol{\epsilon}^{*} - \langle \boldsymbol{\epsilon}^{-1} \rangle_{\Omega}^{-1}$ and $\langle \boldsymbol{\epsilon} \rangle_{\Omega} - \boldsymbol{\epsilon}^{*}$ are nonnegative. Typically, the bounds are quite wide and only provide rough qualitative information. Within the last 50 years, improved estimates have been pursued, with a notable contribution being the Hashin–Shtrikman bounds [39], [40]. There exist several other approaches which seek to estimate or bound the aggregate responses microheterogeneous materials. A complete survey is outside the scope of the present work. For a relatively recent and thorough analysis of a variety of classical approaches, such as the ones briefly mentioned here, see Torquato [100–104] for general interdisciplinary discussions, Jikov et al. [47] for more mathematical aspects, Hashin [39], Mura [72], Nemat–Nasser and Hori [74] for solid mechanics inclined accounts of the subject and Zohdi and Wriggers [128] for computational aspects. Although the primary emphasis in this monograph is on computational methods, we will employ analytical results in order to provide guidance whenever possible. [1] [2]

One goal of this monograph is to blend numerical methods with modeling to the largest extent possible in order to take advantage of the natural synergy between the underlying physics and the solution methods. There is a variety of Computational Electromagnetic Methods which can be lumped into two broad categories: Differential Equation Formulations and Integral Formulations. In the Differential Equation category, the most popular is the Finite Difference Time Domain Method (FDTD), which is also ideally suited to the problems of interest in this monograph, and will be the method of choice.

[1] In the closely related field of solid mechanics, Voigt [106] (1889) is usually cited with the first analysis of the linear effective *mechanical* properties of the microheterogeneous solids, $\langle \boldsymbol{T} \rangle_\Omega = \mathbb{E} : \langle \boldsymbol{\varepsilon} \rangle_\Omega$, where $\boldsymbol{T}$ is the stress and $\boldsymbol{\varepsilon}$ is the (infinitesimal) strain.

[2] Voigt approximated the strain field within an aggregate sample of heterogeneous material as being uniform, leading to $\langle \mathbb{E} \rangle_\Omega$ as an expression of the effective property. Reuss [91] (1929) approximated the stress fields within the aggregate of polycrystalline material as uniform, leading to $\langle \mathbb{E}^{-1} \rangle_\Omega^{-1}$ as the effective property. In 1952, Hill [43] proved that these assumptions provide bounds on the effective property, namely, $\langle \mathbb{E}^{-1} \rangle_\Omega^{-1} \leq \mathbb{E}^* \leq \langle \mathbb{E} \rangle_\Omega$. These inequalities mean that the eigenvalues of the tensors $\mathbb{E}^* - \langle \mathbb{E}^{-1} \rangle_\Omega^{-1}$ and $\langle \mathbb{E} \rangle_\Omega - \mathbb{E}^*$ are nonnegative. Therefore, one can interpret the Voigt and Reuss fields as providing two microfield extremes since the Voigt stress field is one where the tractions at the phase boundaries cannot be in equilibrium (statically inadmissible), while the implied Reuss strains are such that the heterogeneities and the matrix could not be perfectly bonded, i.e., (kinematically inadmissible).

2

Elementary Notation and Mathematical Operations

2.1 Vectors, Products and Norms

In this work, boldface symbols imply vectors or tensors. *A fixed Cartesian coordinate system will be used throughout this monograph.* The unit vectors for such a system are given by the (fixed) mutually orthogonal triad $(\boldsymbol{e}_1, \boldsymbol{e}_2, \boldsymbol{e}_3)$. For the inner product of two vectors, $\boldsymbol{u}$ and $\boldsymbol{v}$, in three dimensions we have

$$\boldsymbol{u} \cdot \boldsymbol{v} = \sum_{i=1}^{3} v_i u_i = u_1 v_1 + u_2 v_2 + u_3 v_3 = ||\boldsymbol{u}|| ||\boldsymbol{v}|| cos\theta, \tag{2.1}$$

where

$$||\boldsymbol{u}|| = \sqrt{u_1^2 + u_2^2 + u_3^2} \tag{2.2}$$

represents the Euclidean norm in $I\!R^3$ and θ is the angle between them. We recall that a norm has three main characteristics for any two bounded vectors $\boldsymbol{u}$ and $\boldsymbol{v}$ ($||\boldsymbol{u}|| < \infty$ and $||\boldsymbol{v}|| < \infty$):

- $||\boldsymbol{u}|| \geq 0$, $||\boldsymbol{u}|| = 0$ if and only if $\boldsymbol{u} = \boldsymbol{0}$,
- $||\boldsymbol{u} + \boldsymbol{v}|| \leq ||\boldsymbol{u}|| + ||\boldsymbol{v}||$ and
- $||\gamma \boldsymbol{u}|| = |\gamma| ||\boldsymbol{u}||$, where γ is a scalar.

Two vectors are said to be orthogonal if $\boldsymbol{u} \cdot \boldsymbol{v} = 0$. The cross (vector) product of two vectors is

$$\boldsymbol{u} \times \boldsymbol{v} = -\boldsymbol{v} \times \boldsymbol{u} = \begin{vmatrix} \boldsymbol{e}_1 & \boldsymbol{e}_2 & \boldsymbol{e}_3 \\ u_1 & u_2 & u_3 \\ v_1 & v_2 & v_3 \end{vmatrix} = ||\boldsymbol{u}|| ||\boldsymbol{v}|| sin\theta \, \boldsymbol{n}, \tag{2.3}$$

where $\boldsymbol{n}$ is the unit normal to the plane formed by the vectors $\boldsymbol{u}$ and $\boldsymbol{v}$.

The temporal differentiation of a vector-valued function is given by

T.I. Zohdi: Electromagnetic Properties of Multiphase Dielectrics, LNACM 64, pp. 5–11.
springerlink.com

$$\frac{d}{dt}\boldsymbol{u}(t) = \frac{du_1(t)}{dt}\boldsymbol{e}_1 + \frac{du_2(t)}{dt}\boldsymbol{e}_2 + \frac{du_3(t)}{dt}\boldsymbol{e}_3 = \dot{u}_1\boldsymbol{e}_1 + \dot{u}_2\boldsymbol{e}_2 + \dot{u}_3\boldsymbol{e}_3. \tag{2.4}$$

The spatial gradient of a scalar-valued function (a dilation to a vector) is given by

$$\nabla_x \phi = \left(\frac{\partial \phi}{\partial x_1}\boldsymbol{e}_1 + \frac{\partial \phi}{\partial x_2}\boldsymbol{e}_2 + \frac{\partial \phi}{\partial x_3}\boldsymbol{e}_3 \right). \tag{2.5}$$

The gradient of a vector-valued function is a direct extension of the preceding definition. For example, $\nabla_x \boldsymbol{u}$ has components of $\frac{\partial u_i}{\partial x_j}$. The divergence of a vector (a contraction to a scalar) is defined by

$$\nabla_x \cdot \boldsymbol{u} = \left(\boldsymbol{e}_1 \frac{\partial}{\partial x_1} + \boldsymbol{e}_2 \frac{\partial}{\partial x_2} + \boldsymbol{e}_3 \frac{\partial}{\partial x_3} \right) \cdot (u_1\boldsymbol{e}_1 + u_2\boldsymbol{e}_2 + u_3\boldsymbol{e}_3) = \left(\frac{\partial u_1}{\partial x_1} + \frac{\partial u_2}{\partial x_2} + \frac{\partial u_3}{\partial x_3} \right). \tag{2.6}$$

The curl of a vector is defined as

$$\nabla_x \times \boldsymbol{u} = \begin{vmatrix} \boldsymbol{e}_1 & \boldsymbol{e}_2 & \boldsymbol{e}_3 \\ \frac{\partial}{\partial x_1} & \frac{\partial}{\partial x_2} & \frac{\partial}{\partial x_3} \\ u_1 & u_2 & u_3 \end{vmatrix}. \tag{2.7}$$

The triple product of three vectors is

$$\boldsymbol{w} \cdot (\boldsymbol{u} \times \boldsymbol{v}) = \left(\begin{vmatrix} w_1 & w_2 & w_3 \\ u_1 & u_2 & u_3 \\ v_1 & v_2 & v_3 \end{vmatrix} \right) = (\boldsymbol{w} \times \boldsymbol{u}) \cdot \boldsymbol{v}. \tag{2.8}$$

This represents the volume of a parallelepiped formed by the three vectors.

2.2 Basic Linear Algebra

If we consider the second order tensor $\boldsymbol{A}$ with its matrix representation

$$[\boldsymbol{A}] \stackrel{\text{def}}{=} \begin{bmatrix} A_{11} & A_{12} & A_{13} \\ A_{21} & A_{22} & A_{23} \\ A_{31} & A_{32} & A_{33} \end{bmatrix}, \tag{2.9}$$

the matrix $[\boldsymbol{A}]$ is said to be symmetric if $[\boldsymbol{A}] = [\boldsymbol{A}]^T$ and skew-symmetric if $[\boldsymbol{A}] = -[\boldsymbol{A}]^T$. A first order contraction (inner product) of two matrices is defined by

$$\boldsymbol{A} \cdot \boldsymbol{B} = [\boldsymbol{A}][\boldsymbol{B}] \text{ which has components of } \sum_{j=1}^{N} A_{ij}B_{jk} = C_{ik} \tag{2.10}$$

where it is clear that the range of the inner index j must be the same for $[\boldsymbol{A}]$ and $[\boldsymbol{B}]$. The second order inner product of two matrices is

$$\boldsymbol{A} : \boldsymbol{B} = A_{ij}B_{ij} = tr([\boldsymbol{A}]^T[\boldsymbol{B}]) \tag{2.11}$$

Some properties of the determinant are (where, for example, $[\boldsymbol{A}]$ is a 3×3 matrix):

- $det[\boldsymbol{A}] = A_{11}(A_{22}A_{33} - A_{32}A_{23}) - A_{12}(A_{21}A_{33} - A_{31}A_{23}) + A_{13}(A_{21}A_{32} - A_{31}A_{22})$,
- $det[\boldsymbol{I}] = 1$, $det\, \alpha[\boldsymbol{A}] = \alpha^3 det\, [\boldsymbol{A}]$, where α is a scalar,
- $det[\boldsymbol{A}][\boldsymbol{B}] = det[\boldsymbol{A}]det[\boldsymbol{B}]$, $det[\boldsymbol{A}]^T = det[\boldsymbol{A}]$ and $det[\boldsymbol{A}]^{-1} = \frac{1}{det[\boldsymbol{A}]}$.

An important use of the determinant is in forming the inverse by

$$[\boldsymbol{A}]^{-1} = \frac{adj[\boldsymbol{A}]}{det[\boldsymbol{A}]}, \qquad adj[\boldsymbol{A}] \stackrel{\text{def}}{=} \begin{bmatrix} C_{11} & C_{12} & C_{13} \\ C_{21} & C_{22} & C_{23} \\ C_{31} & C_{32} & C_{33} \end{bmatrix}^T, \tag{2.12}$$

where the so-called cofactors are

$$\boxed{\begin{array}{ll} C_{11} = A_{22}A_{33} - A_{32}A_{23} & C_{12} = -(A_{21}A_{33} - A_{31}A_{23}) \\ C_{13} = A_{21}A_{32} - A_{31}A_{22} & C_{21} = -(A_{12}A_{33} - A_{32}A_{13}) \\ C_{22} = A_{11}A_{33} - A_{31}A_{13} & C_{23} = -(A_{11}A_{32} - A_{31}A_{12}) \\ C_{31} = A_{12}A_{23} - A_{22}A_{13} & C_{32} = -(A_{11}A_{23} - A_{21}A_{13}). \\ C_{33} = A_{11}A_{22} - A_{21}A_{12} & \end{array}} \tag{2.13}$$

The rule of transposes for two $n \times n$ matrices

$$([\boldsymbol{A}][\boldsymbol{B}])^T = [\boldsymbol{B}]^T[\boldsymbol{A}]^T. \tag{2.14}$$

The rule of inverses for two invertible $n \times n$ matrices is

$$([\boldsymbol{A}][\boldsymbol{B}])^{-1} = [\boldsymbol{B}]^{-1}[\boldsymbol{A}]^{-1} \qquad \text{and} \qquad [\boldsymbol{A}]^{-1}[\boldsymbol{A}] = [\boldsymbol{A}][\boldsymbol{A}]^{-1} = [\boldsymbol{I}] \tag{2.15}$$

where $[\boldsymbol{I}]$ is the identity matrix. Clearly, $[\boldsymbol{A}]^{-1}$ exists only when $det[\boldsymbol{A}] \neq 0$.

The mathematical definitions of an eigenvalue, a scalar denoted Λ and eigenvector, a vector denoted $\boldsymbol{\mathcal{E}}$, of a matrix $[\boldsymbol{A}]$ are

$$[\boldsymbol{A}]\{\boldsymbol{\mathcal{E}}\} = \Lambda\{\boldsymbol{\mathcal{E}}\}. \tag{2.16}$$

We note that for any given tensor $\boldsymbol{A}$ of order 2 (a 3×3 matrix), if we set the determinant $det[\boldsymbol{A} - \Lambda\boldsymbol{I}] = 0$, it can be shown that the so-called "characteristic" polynomial is

$$det(\boldsymbol{A} - \Lambda \boldsymbol{I}) = -\Lambda^3 + I_A \Lambda^2 - II_A \Lambda + III_A = 0, \tag{2.17}$$

where

$$\boxed{\begin{aligned} I_A &= tr(\boldsymbol{A}) = \Lambda_1 + \Lambda_2 + \Lambda_3 \\ II_A &= \tfrac{1}{2}((tr(\boldsymbol{A}))^2 - tr(\boldsymbol{A}^2)) = \Lambda_1\Lambda_2 + \Lambda_2\Lambda_3 + \Lambda_1\Lambda_3 \\ III_A &= det(\boldsymbol{A}) = \tfrac{1}{6}((tr\boldsymbol{A})^3 + 2tr\boldsymbol{A}^3 - 3(tr\boldsymbol{A}^2)(tr\boldsymbol{A})) = \Lambda_1\Lambda_2\Lambda_3. \end{aligned}} \tag{2.18}$$

Since I_A, II_A and III_A can be written in terms of $tr\boldsymbol{A}$, which is invariant under frame rotation, they too are invariant under frame rotation. The main properties to remember about eigenvalues and eigenvectors are:

1. If $[\boldsymbol{A}]$ $(n \times n)$ has n linearly independent eigenvectors, then it is diagonalizable by a matrix formed by columns of the eigenvectors, for example, for a 3×3 matrix

$$\begin{bmatrix} \Lambda_1 & 0 & 0 \\ 0 & \Lambda_2 & 0 \\ 0 & 0 & \Lambda_3 \end{bmatrix} = \begin{bmatrix} \mathcal{E}_1^{(1)} & \mathcal{E}_1^{(2)} & \mathcal{E}_1^{(3)} \\ \mathcal{E}_2^{(1)} & \mathcal{E}_2^{(2)} & \mathcal{E}_2^{(3)} \\ \mathcal{E}_3^{(1)} & \mathcal{E}_3^{(2)} & \mathcal{E}_3^{(3)} \end{bmatrix}^{-1} \begin{bmatrix} A_{11} & A_{12} & A_{13} \\ A_{21} & A_{22} & A_{23} \\ A_{31} & A_{32} & A_{33} \end{bmatrix} \begin{bmatrix} \mathcal{E}_1^{(1)} & \mathcal{E}_1^{(2)} & \mathcal{E}_1^{(3)} \\ \mathcal{E}_2^{(1)} & \mathcal{E}_2^{(2)} & \mathcal{E}_2^{(3)} \\ \mathcal{E}_3^{(1)} & \mathcal{E}_3^{(2)} & \mathcal{E}_3^{(3)} \end{bmatrix} \tag{2.19}$$

2. If $[\boldsymbol{A}]$ $(n \times n)$ has n distinct eigenvalues, then the eigenvectors are linearly independent
3. If $[\boldsymbol{A}]$ $(n \times n)$ is symmetric, then its eigenvalues are real. If the eigenvalues are distinct, then the eigenvectors are orthogonal.

A quadratic form is such that

$$\{\boldsymbol{x}\}^T[\boldsymbol{A}]\{\boldsymbol{x}\} \stackrel{\text{def}}{=} \begin{bmatrix} x_1 & x_2 & x_3 \end{bmatrix} \begin{bmatrix} A_{11} & A_{12} & A_{13} \\ A_{21} & A_{22} & A_{23} \\ A_{31} & A_{32} & A_{33} \end{bmatrix} \begin{bmatrix} x_1 \\ x_2 \\ x_3 \end{bmatrix}. \tag{2.20}$$

A matrix $[\boldsymbol{A}]$ is said to be positive definite if the quadratic form is positive for all nonzero vectors $\boldsymbol{x}$. Clearly, from equation 2.19, a positive definite matrix must have positive eigenvalues.

To perform a coordinate transform for a 3×3 matrix $[\boldsymbol{A}]$ from one Cartesian coordinate system to another (denoted with a $\hat{(\cdot)}$), we apply a transformation matrix $[\boldsymbol{Q}]$:

$$[\hat{\boldsymbol{A}}] = [\boldsymbol{Q}][\boldsymbol{A}][\boldsymbol{Q}]^{-1}. \tag{2.21}$$

If $\boldsymbol{Q}$ is an orthogonal matrix, then $\boldsymbol{Q}^{-1} = \boldsymbol{Q}^T$ (denoted "unitary").

The standard axes rotators (Figure 2.1) about the x_1 axis are

$$Rot(x_1) \stackrel{\text{def}}{=} \begin{bmatrix} 1 & 0 & 0 \\ 0 & cos\theta_1 & sin\theta_1 \\ 0 & -sin\theta_1 & cos\theta_1 \end{bmatrix}, \tag{2.22}$$

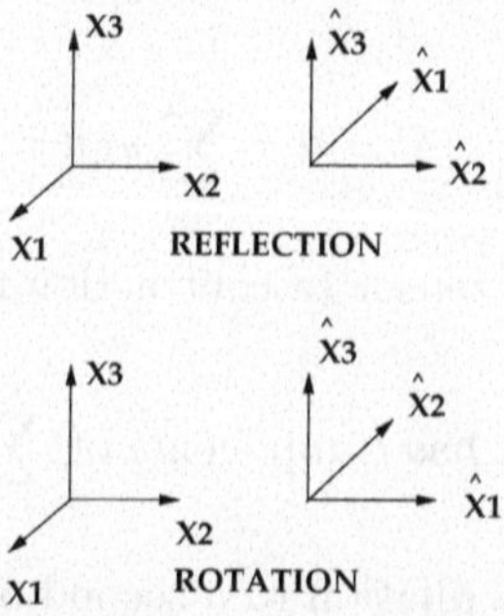

Fig. 2.1. Top: reflection with respect to the $x_2 - x_3$ plane. Bottom: rotation with respect to the x_3 axis.

about the x_2 axis

$$Rot(x_2) \stackrel{\text{def}}{=} \begin{bmatrix} cos\theta_2 & 0 & -sin\theta_2 \\ 0 & 1 & 0 \\ sin\theta_2 & 0 & cos\theta_2 \end{bmatrix}, \tag{2.23}$$

and about the x_3 axis

$$Rot(x_3) \stackrel{\text{def}}{=} \begin{bmatrix} cos\theta_3 & sin\theta_3 & 0 \\ -sin\theta_3 & cos\theta_3 & 0 \\ 0 & 0 & 1 \end{bmatrix}. \tag{2.24}$$

The standard axes reflectors are, with respect to the $x_2 - x_3$ plane,

$$Ref(x_1) \stackrel{\text{def}}{=} \begin{bmatrix} -1 & 0 & 0 \\ 0 & 1 & 0 \\ 0 & 0 & 1 \end{bmatrix}, \tag{2.25}$$

with respect to the $x_1 - x_3$ plane,

$$Ref(x_2) \stackrel{\text{def}}{=} \begin{bmatrix} 1 & 0 & 0 \\ 0 & -1 & 0 \\ 0 & 0 & 1 \end{bmatrix}, \tag{2.26}$$

and with respect to the $x_1 - x_2$ plane,

$$Ref(x_3) \stackrel{\text{def}}{=} \begin{bmatrix} 1 & 0 & 0 \\ 0 & 1 & 0 \\ 0 & 0 & -1 \end{bmatrix}. \tag{2.27}$$

2.3 Integral Transformations

The divergence of a vector-valued function (a contraction to a scalar-valued function) is defined by

$$\nabla_x \cdot \boldsymbol{u} = \sum_{i=1}^{N} u_{i,i}, \tag{2.28}$$

whereas for a second order tensor (a contraction to a vector)

$$\nabla_x \cdot \boldsymbol{A} \text{ has components of } \sum_{j=1}^{N} A_{ij,j}. \tag{2.29}$$

The gradient of a vector (a dilation to a second order tensor) is

$$\nabla_x \boldsymbol{u} \text{ has components of } u_{i,j}, \tag{2.30}$$

whereas for a second order tensor (a dilation to a third order tensor)

$$\nabla_x \boldsymbol{A} \text{ has components of } A_{ij,k}. \tag{2.31}$$

The gradient of a scalar (a dilation to a vector)

$$\nabla_x \phi \text{ has components of } \phi_{,i}. \tag{2.32}$$

The scalar product of two second order tensors, for example, the gradients of first order vectors, is defined as

$$\nabla_x \boldsymbol{v} : \nabla_x \boldsymbol{u} = \underbrace{\frac{\partial v_i}{\partial x_j}\frac{\partial u_i}{\partial x_j}}_{\text{in Cartesian bases}} \stackrel{\text{def}}{=} v_{i,j}u_{i,j} \qquad i,j = 1,2,3, \tag{2.33}$$

where $\partial u_i/\partial x_j, \partial v_i/\partial x_j$ are partial derivatives of u_i and v_i, and where u_i, v_i are the Cartesian components of $\boldsymbol{u}$ and $\boldsymbol{v}$, and

$$\nabla_x \boldsymbol{u} \cdot \boldsymbol{n} \text{ has components of } \underbrace{u_{i,j}n_j}_{\text{in Cartesian bases}}, \qquad i,j = 1,2,3. \tag{2.34}$$

For a scalar, we have

$$\int_{\Omega} \nabla_x \phi \, d\Omega = \int_{\partial\Omega} \phi \boldsymbol{n} \, dA \qquad \int_{\Omega} \phi_{,i} \, d\Omega = \int_{\partial\Omega} \phi n_i \, dA, \tag{2.35}$$

and for a vector,

$$\int_{\Omega} \nabla_x \boldsymbol{u} \, d\Omega = \int_{\partial\Omega} \boldsymbol{u} \otimes \boldsymbol{n} \, dA \qquad \int_{\Omega} u_{i,j} \, d\Omega = \int_{\partial\Omega} u_i n_j \, dA. \tag{2.36}$$

The divergence theorem for vectors is

$$\int_{\Omega} \nabla_x \cdot \boldsymbol{u} \, d\Omega = \int_{\partial\Omega} \boldsymbol{u} \cdot \boldsymbol{n} \, dA \qquad \int_{\Omega} u_{i,i} \, d\Omega = \int_{\partial\Omega} u_i n_i \, dA, \tag{2.37}$$

and analogously for a tensor $\boldsymbol{B}$

$$\int_{\Omega} \nabla_x \cdot \boldsymbol{B} \, d\Omega = \int_{\partial\Omega} \boldsymbol{B} \cdot \boldsymbol{n} \, dA \qquad \int_{\Omega} B_{ij,j} \, d\Omega = \int_{\partial\Omega} B_{ij} n_j \, dA, \tag{2.38}$$

where $\boldsymbol{n}$ is the outward normal to the bounding surface. These standard operations arise throughout the analysis. A generalization of these last results is

$$\int_{\Omega} \nabla_x * \boldsymbol{B} \, d\Omega = \int_{\partial\Omega} \boldsymbol{n} * \boldsymbol{B} \, dA, \tag{2.39}$$

where when $* = \cdot$, we have the divergence theorem and when $* = \times$, we have the "cross-product" theorem. [1] For proofs, see Chandrasekharaiah and Debnath [10], or Malvern [64].

[1] Also, we have the pointwise product rule:

$$\frac{d}{dt}(\boldsymbol{a} * \boldsymbol{b}) = \frac{d\boldsymbol{a}}{dt} * \boldsymbol{b} + \boldsymbol{a} * \frac{d\boldsymbol{a}}{dt}. \tag{2.40}$$

3

Governing Electromagnetics: Maxwell's Equations

Some fundamental definitions and observations in conjunction with electromagnetic phenomena are:

- If a point charge q experiences a force $\boldsymbol{\Psi}^e$, the electric field, $\boldsymbol{E}$, at a position of the charge is defined by $\boldsymbol{\Psi}^e = q\boldsymbol{E}$.
- If the charge is moving, another force may arise, $\boldsymbol{\Psi}^m$, which is proportional to its velocity $\boldsymbol{v}$. This other field is denoted as the "magnetic induction" (induced force field) or just the "magnetic field," $\boldsymbol{B}$, such that $\boldsymbol{\Psi}^m = q\boldsymbol{v} \times \boldsymbol{B}$.
- If the forces occur concurrently (the charge is moving through the region possessing both electric and magnetic fields), then $\boldsymbol{\Psi}^{em} = q\boldsymbol{E} + q\boldsymbol{v} \times \boldsymbol{B}$.
- Electric fields are produced by both electric charges and time-varying magnetic fields. Similarly, magnetic fields are induced by electric currents and by time-varying electric fields.

From these elementary definitions and observations, Maxwell's equations follow.

3.1 Coulomb Forces

Coulomb's law resulted as an outgrowth of various attempts to describe the forces that arise between small charged bodies. *It is remarkably simple, and forms the basis of electromagnetics and ultimately, Maxwell's equations.* The force, due solely to (static) electrical effects, between two electrically charged particles is given by (Figure 3.1)

$$\boldsymbol{\Psi}^e = \frac{q_1 q_2}{4\pi\epsilon_o d_{12}^2} \boldsymbol{e}_{21}, \tag{3.1}$$

where $\boldsymbol{\Psi}^e$ is the force acting between the particles (Newtons), q_1 is the charge of particle 1 (Coulombs), q_2 is the charge of particle 2, $\epsilon_o = \frac{10^{-9}}{36\pi} = 8.854 \times$

T.I. Zohdi: Electromagnetic Properties of Multiphase Dielectrics, LNACM 64, pp. 13–37.
springerlink.com

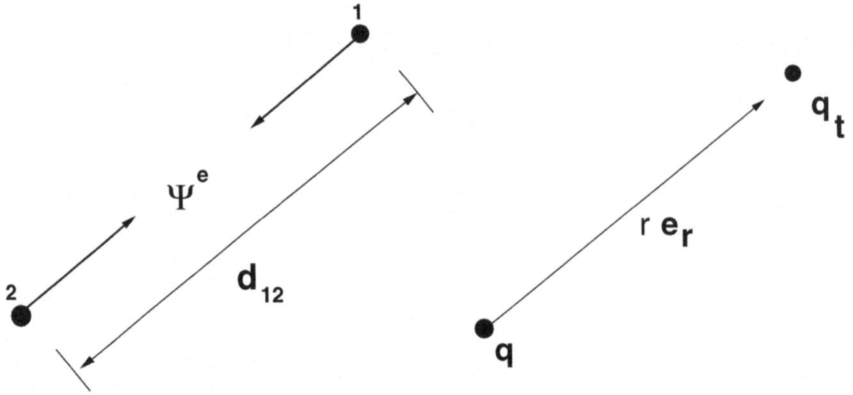

Fig. 3.1. Left: Two charged point masses with the force that acts between them. Right: A test charge with (relative) position vector $r\boldsymbol{e}_r$ pointing from q to q_t.

10^{-12} $farads/m$ is the free-space (vacuum) permittivity, $\boldsymbol{e}_{21}$ is the unit vector between the particles' centers (from 2 to 1) and d_{12} is the separation distance. For multiple charges,

$$\boldsymbol{\Psi}^e = \sum_{k=2}^{N} \frac{q_1 q_k}{4\pi\epsilon_o d_{1k}^2} \boldsymbol{e}_{k1}. \tag{3.2}$$

For other media, it is customary to write $\epsilon = \epsilon_o \epsilon_r$, where ϵ_r is the relative permittivity or "dielectric" constant. For point charges of like sign, the Coulombic force is one of repulsion, while for unlike charges, the force is attractive. This simple model serves as the starting point for constructing the classical continuum theory of electrodynamics. There are numerous books on the subject, and we attempt here to only give a relatively brief introduction to the essentials.

Remark: In order to make this introductory chapter as clear as possible, we employ the following notation:

- Ω is the domain under consideration,
- V_Ω is the volume of the domain under consideration,
- A_Ω is the surface area of the domain under consideration,
- C_Ω is a contour of the domain under consideration.

3.2 Electric Field

The force field in the region of an isolated charge q is spherically symmetric. For example, consider the (relative) position vector, $r\boldsymbol{e}_r$, from q to q_t (a test charge (Figure 3.1)), with a corresponding force field on the test charge given by

$$\boldsymbol{\Psi}_t^e = \frac{q_t q}{4\pi\epsilon_o r^2}\boldsymbol{e}_r. \qquad (3.3)$$

The electric field intensity $\boldsymbol{E}$, due to q, is defined as the force per unit charge on q_t, that is,

$$\boldsymbol{E} = \frac{\boldsymbol{\Psi}_t^e}{q_t} = \frac{q}{4\pi\epsilon_o r^2}\boldsymbol{e}_r. \qquad (3.4)$$

When the charge (q) is distributed throughout the volume, we have ($\mathcal{P}$)

$$\mathcal{P} = \frac{dq}{dV} \Rightarrow \mathcal{P}dV = dq, \qquad (3.5)$$

and therefore, from a distributed "cloud" of charges, one can represent the electric field at a point p as (Figure 3.2)

$$\boldsymbol{E}_p = \int_{V_\Omega} \frac{\mathcal{P}}{4\pi\epsilon_o r^2}\boldsymbol{e}_r\, dV, \qquad (3.6)$$

where the (relative) position vector $r\boldsymbol{e}_r$ points from each distributed (differential) charge to the point p.

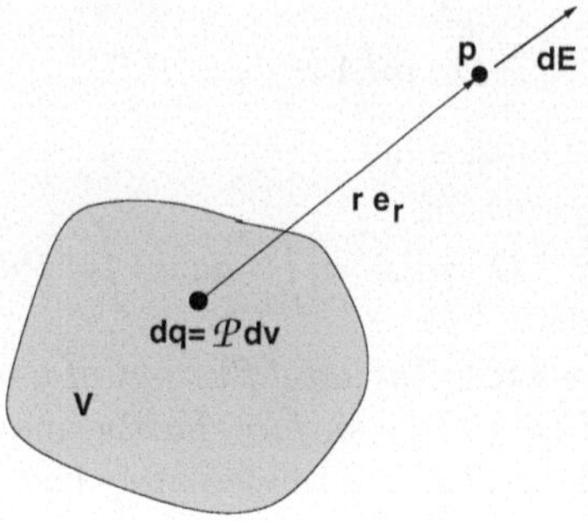

Fig. 3.2. A cloud of distributed charges and their effect on a point p.

3.3 Electric Flux

Since $dq = \mathcal{P}dV$, we may write (the total charge in V_Ω)

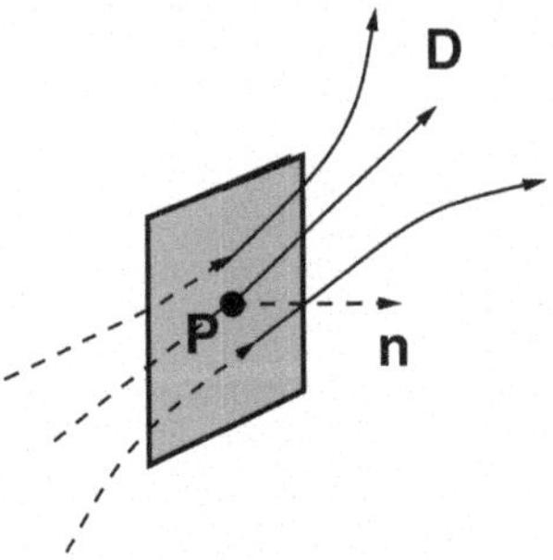

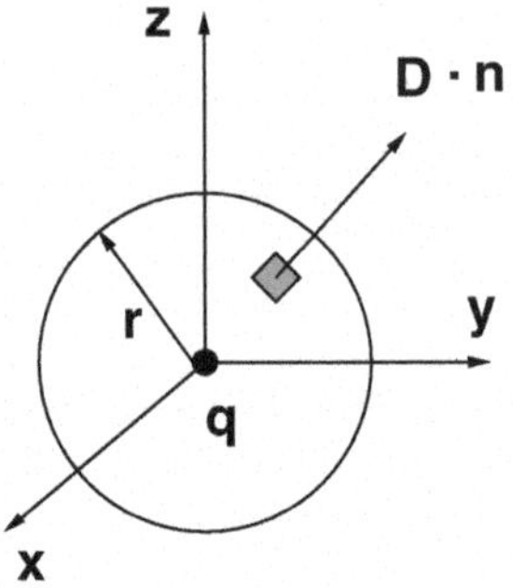

Fig. 3.3. Left: The electric flux acting at a point p with a surface normal $\boldsymbol{n}$. Right: A point charge and the induced normal electric field flux.

$$q = \int_{V_\Omega} \mathcal{P} dV. \tag{3.7}$$

We denote the electric flux, $\mathcal{F}$, as a scalar field as originating on a positive charge and terminating at a negative charge. If there is no negative charge, then $\mathcal{F}$ terminates at infinity. We denote the electric flux density by $\boldsymbol{D}$, a vector field. By definition, one Coulomb of electric charge gives rise to one Coulomb of electric flux. Referring to Figure 3.3, we have the electric flux density, $\boldsymbol{D}$, defined by

$$\boldsymbol{D} \cdot \boldsymbol{n} dA = dq = \mathcal{P} dV, \tag{3.8}$$

and we directly obtain Gauss's law:

$$\int_{A_\Omega} \boldsymbol{D} \cdot \boldsymbol{n} dA = \int_{V_\Omega} dq = \int_{V_\Omega} \mathcal{P} dV. \tag{3.9}$$

Explicitly, Gauss's law states: *The total flux out of a closed surface is equal to the net charge enclosed within the surface.* Furthermore, if we consider an isolated point charge q (Figure 3.3), we immediately have an induced spherically-symmetric field

$$q = \int_{A_\Omega} \boldsymbol{D} \cdot \boldsymbol{n}\, dA = D \int_{A_\Omega} dA = D 4\pi r^2, \tag{3.10}$$

where at a fixed distance r, $\boldsymbol{D}$ has a constant magnitude D. Therefore, from Coulomb's law,

$$D = \frac{q}{4\pi r^2} = \epsilon_o E \Rightarrow D = \epsilon_o E. \tag{3.11}$$

The previous result holds for a vacuum, but motivates constitutive laws for nonvacuum media introduced later.

3.4 Current Density and Conductors

Electric current is the rate of transport of electric charge past a specified point or surface. The unit of current is Ampere (1A=1C/s). The current density is denoted $\boldsymbol{J}$ (A/m^2). The total current is

$$I = \int_{A_\Omega} \boldsymbol{J} \cdot \boldsymbol{n} \, dA. \tag{3.12}$$

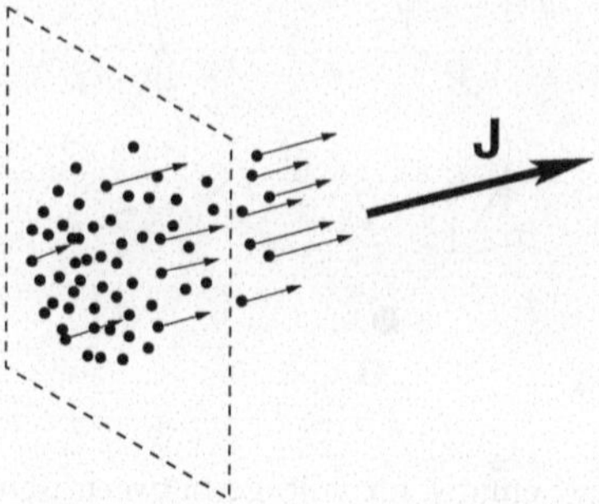

Fig. 3.4. Current created by charge flow.

In theory, if the charge carriers would move in an unimpeded fashion, then force on the charge carriers $\mathcal{F} = q\boldsymbol{E}$ would produce a constant acceleration (if there were negligible magnetic fields). However, when in a medium, collisions slow down the charge carriers. On average, the random velocity components, due to random collisions, cancel one another out and we are left with a group "drift velocity" (average velocity of the charge carriers), $\boldsymbol{\mathcal{V}}$, which is proportional to the electric field, as well as the induced effects from magnetic field

$$\boldsymbol{\mathcal{V}} \propto k(\boldsymbol{E} + \boldsymbol{v} \times \boldsymbol{B}), \tag{3.13}$$

where k is a mobility constant and $\boldsymbol{v}$ is the velocity of the medium (continuum). One can write

$$\boldsymbol{J} = \mathcal{P}\mathcal{V} \stackrel{\text{def}}{=} \boldsymbol{\sigma} \cdot (\boldsymbol{E} + \boldsymbol{v} \times \boldsymbol{B}), \tag{3.14}$$

where $\boldsymbol{\sigma}$ is the conductivity in Siemens/m, which is inversely proportional to the resistivity. With increasing temperature, the vibratory motion and mutual collisions of ions increase, leading to a reduction of the conductivity with temperature.

Remark: In the absence of any electromagnetic field, the group drift velocity is zero (individual charge carriers have a completely random velocity).

3.5 Work and Voltage

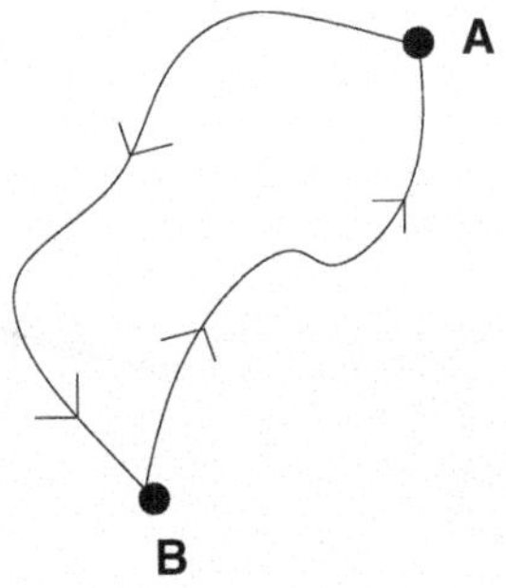

Fig. 3.5. A schematic for voltage between points A and B.

The work done by moving a point charge from location B to location A in a *static electric field* is independent of the path taken since

$$\int_A^B q\boldsymbol{E} \cdot d\boldsymbol{L} = -\int_B^A q\boldsymbol{E} \cdot d\boldsymbol{L} \Rightarrow \oint_A^B \boldsymbol{E} \cdot d\boldsymbol{L} = 0. \tag{3.15}$$

The potential of a point A with respect to a point B is defined as the work done in moving a unit positive charge q_u from B to A:

$$voltage \stackrel{\text{def}}{=} \Pi_{A-B} = \frac{W}{q_u} = -\int_B^A \boldsymbol{E} \cdot d\boldsymbol{L}. \tag{3.16}$$

Π_{A-B} (J/C = *volt*) can be considered as the potential difference between points A and B or *voltage*. When Π_{A-B} is positive, work must be expended to move the unit positive charge from B to A, and point A is said to have a higher potential than point B. Now, consider that

$$d\Pi = \frac{\partial \Pi}{\partial x}dx + \frac{\partial \Pi}{\partial y}dy + \frac{\partial \Pi}{\partial z}dz, \tag{3.17}$$

and

$$\nabla_x \Pi = \frac{\partial \Pi}{\partial x}\boldsymbol{e}_x + \frac{\partial \Pi}{\partial y}\boldsymbol{e}_y + \frac{\partial \Pi}{\partial z}\boldsymbol{e}_z, \tag{3.18}$$

and

$$d\boldsymbol{L} = dx\boldsymbol{e}_x + dy\boldsymbol{e}_y + dz\boldsymbol{e}_z, \tag{3.19}$$

therefore,

$$d\Pi = \nabla_x \Pi \cdot d\boldsymbol{L} = -\boldsymbol{E} \cdot d\boldsymbol{L}. \tag{3.20}$$

Thus,

$$\boldsymbol{E} = -\nabla_x \Pi. \tag{3.21}$$

3.6 Maxwell–Ampere Law and the Magnetic Field

Ampere's law states that the line integral component of the magnetic field strength around a closed path is equal to the current enclosed by the path (Figure 3.6):

$$\int_{C_\Omega} \boldsymbol{H} \cdot d\boldsymbol{L} = I_{\text{enclosed}} = \int_{A_\Omega} \boldsymbol{J} \cdot \boldsymbol{n}\, dA. \tag{3.22}$$

For example, take a circular path around a fixed straight wire (Figure 3.7), that is,

$$\int_{C_\Omega} \boldsymbol{H} \cdot d\boldsymbol{L} = \underbrace{H2\pi R}_{\text{for straight wire}} = I \Rightarrow H = \frac{I}{2\pi R}. \tag{3.23}$$

Using Stokes's theorem, and because one may write $I_{\text{enclosed}} = \int_{A_\Omega} \boldsymbol{J} \cdot \boldsymbol{n}\, dA$ (a fixed domain),

$$\int_{C_\Omega} \boldsymbol{H} \cdot d\boldsymbol{L} = \int_{A_\Omega} \nabla_x \times \boldsymbol{H} \cdot \boldsymbol{n}\, dA = I_{\text{enclosed}} = \int_{A_\Omega} \boldsymbol{J} \cdot \boldsymbol{n}\, dA. \tag{3.24}$$

Since A is arbitrary, we have

$$\nabla_x \times \boldsymbol{H} = \boldsymbol{J}. \tag{3.25}$$

Like $\boldsymbol{D}$, the magnetic field strength $\boldsymbol{H}$ depends only on moving charges and is independent of the medium. The force field associated with $\boldsymbol{H}$ is the magnetic field flux density ($\boldsymbol{B}$) given by, for linear, isotropic, materials

$$\boldsymbol{B} = \mu \boldsymbol{H}, \tag{3.26}$$

where $\mu = \mu_o \mu_r$ is the permeability of the medium. The units of $\boldsymbol{B}$ are Teslas, 1 T=1 N/A-m, where $\mu_o = 4\pi \times 10^{-7}$ Henries/m. μ_r is a number that is typically near unity, except for ferromagnetic materials. The magnetic flux, Φ (in Webers, Wb), through a surface is defined as[1]

$$\Phi = \int_{A_\Omega} \boldsymbol{B} \cdot \boldsymbol{n} \, dA. \tag{3.27}$$

The lines of magnetic flux are closed curves with no starting point or point of termination. This is in contrast with the electric flux, $\mathcal{F}$, which originates on a positive charge and terminates on a negative charge. All magnetic flux Φ that enter a closed surface must leave the surface. Thus, $\boldsymbol{B}$ fields have no sources or sinks, which is mathematically expressed by

$$\nabla_x \cdot \boldsymbol{B} = 0. \tag{3.28}$$

Unlike electrical charges, no isolated magnetic "charges" ("monopoles") have ever been found. Therefore, the magnetic flux can be written as

$$\Phi = \int_{A_\Omega} \boldsymbol{B} \cdot \boldsymbol{n} \, dA = 0 \tag{3.29}$$

since the magnetic field lines form closed loops.[2] Thus, we have (Figure 3.6)

$$\int_{C_\Omega} \boldsymbol{H} \cdot d\boldsymbol{L} = \int_{A_\Omega} (\nabla_x \times \boldsymbol{H}) \cdot \boldsymbol{n} \, dA = I_{\text{enclosed}}. \tag{3.30}$$

Maxwell's equations first assumption (modification to Ampere's law) was

$$\int_{C_\Omega} \boldsymbol{H} \cdot d\boldsymbol{L} = \int_{A_\Omega} \nabla_x \times \boldsymbol{H} \cdot \boldsymbol{n} \, dA = I_{\text{enclosed}} = \int_{A_\Omega} (\boldsymbol{J}^{ext} + \boldsymbol{J}_c + \frac{\partial \boldsymbol{D}}{\partial t} \cdot \boldsymbol{n}) \, dA, \tag{3.31}$$

where $\boldsymbol{J}^{ext}$ is due to current sources, $\boldsymbol{J}_c$ is due to conductive current.

Remarks: A differential magnetic field strength, $d\boldsymbol{H}$, results from a differential current element $Id\boldsymbol{L}$. The field varies inversely with distance squared, is independent of the surrounding medium, and has a direction given by the cross-product of $Id\boldsymbol{L}$ and $\boldsymbol{e}_r$, known as the Biot–Savart law (Figure 3.7):

---

[1] Note that $1T = 1Wb/m^2$ and $1H = 1Wb/A$.

[2] Later, we will introduce phenomenological terms that produce an effect that leads to $\nabla_x \cdot \boldsymbol{B} \neq 0$.

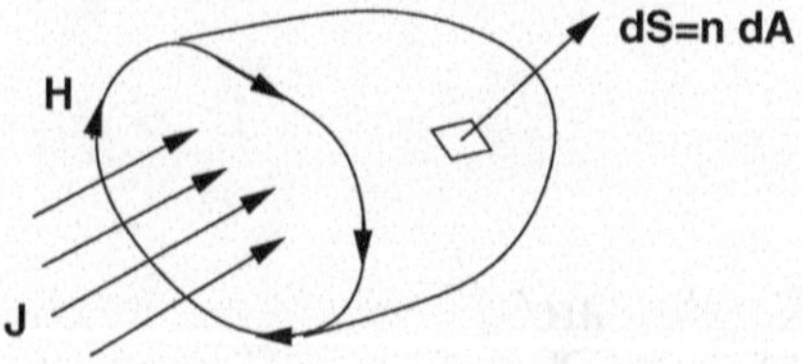

Fig. 3.6. Ampere's law: a pictorial representation.

$$d\boldsymbol{H} = \frac{Id\boldsymbol{L} \times \boldsymbol{e}_r}{4\pi R^2}, \tag{3.32}$$

where $R = ||\boldsymbol{r}||$ and $\boldsymbol{e}_r = \frac{\boldsymbol{r}}{||\boldsymbol{r}||}$, which points from the current element to the point at which $\boldsymbol{H}$ is to be determined, for example, p. In the standard theory, current elements have no separate existence; all elements making up the complete current filament contribute to $\boldsymbol{H}$ and must be included. Thus, the magnetic field acting at a point p

$$\boldsymbol{H} = \int_{C_\Omega} \frac{Id\boldsymbol{L} \times \boldsymbol{e}_r}{4\pi R^2}. \tag{3.33}$$

3.7 Faraday's Induction Law

In approximately 1831, Faraday observed that changing a magnetic field generated a current. By thrusting a magnet into a coil, a voltage was induced; a so-called "induced electromotive force" (EMF). Three observations are important:

- The induced EMF depends on the rate of change of $\boldsymbol{B}$.
- The induced EMF is proportional to the area of a (closed) loop penetrated perpendicularly by the field (normal flux).
- When the field is constant, the induced EMF is proportional to the rate of change of the perpendicular area penetrated.

We define the differential flux of the magnetic field through the wire loop by

$$d\Psi^m = \boldsymbol{B} \cdot \boldsymbol{n}\, dA. \tag{3.34}$$

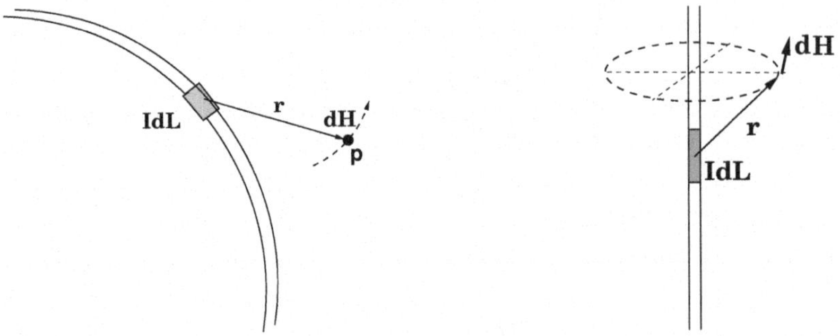

Fig. 3.7. Left: Induced magnetic field at a point p. Right: The magnetic field induced by current flowing in a straight wire.

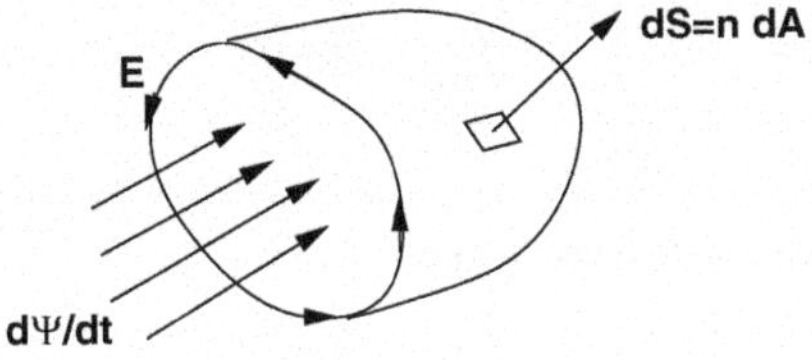

Fig. 3.8. The flow of a magnetic field.

Thus, the total flux is

$$\boldsymbol{\Psi}^m = \int_{A_\Omega} \boldsymbol{B} \cdot \boldsymbol{n} \, dA, \tag{3.35}$$

and the induced EMF is

$$EMF = -\frac{d\boldsymbol{\Psi}^m}{dt}. \tag{3.36}$$

The sum of the EMF around the contour is therefore

$$\int_{C_\Omega} (\boldsymbol{E} + \boldsymbol{v} \times \boldsymbol{B}) \cdot d\boldsymbol{L} = -\frac{d}{dt} \int_{A_\Omega} \boldsymbol{B} \cdot \boldsymbol{n}\, dA. \tag{3.37}$$

The right-hand side can be decomposed into two pieces:

$$\int_{C_\Omega} (\boldsymbol{E} + \boldsymbol{v} \times \boldsymbol{B}) \cdot d\boldsymbol{L} = -\int_{A_\Omega} \frac{\partial \boldsymbol{B}}{\partial t} \cdot \boldsymbol{n}\, dA + \int_{C_\Omega} \boldsymbol{v} \times \boldsymbol{B} \cdot d\boldsymbol{L}, \tag{3.38}$$

and therefore Maxwell–Faraday's law is

$$\int_{C_\Omega} \boldsymbol{E} \cdot d\boldsymbol{L} = -\int_{A_\Omega} \frac{\partial \boldsymbol{B}}{\partial t} \cdot \boldsymbol{n}\, dA. \tag{3.39}$$

Remark 1: One may write

$$\int_{C_\Omega} \boldsymbol{E} \cdot d\boldsymbol{L} = \int_{A_\Omega} (\nabla_x \times \boldsymbol{E}) \cdot \boldsymbol{n}\, dA = -\int_{A_\Omega} (\boldsymbol{M}^{ext} + \boldsymbol{M}_c + \frac{\partial \boldsymbol{B}}{\partial t}) \cdot \boldsymbol{n}\, dA, \tag{3.40}$$

where $\boldsymbol{M}^{ext}$ and $\boldsymbol{M}_c$ are phenomenological terms that frequently appear in the literature to account for magnetic "sources" and "magnetic conduction." We shall keep these terms throughout the formulations, but with an implicit warning that they are difficult to justify from first principles.

Remark 2: The result in Equation 3.38 can be derived directly from use of the transport theorem, along with the identity

$$\nabla_x \times (\boldsymbol{B} \times \boldsymbol{v}) = \boldsymbol{B}(\nabla_x \cdot \boldsymbol{v}) - \boldsymbol{v}(\nabla_x \cdot \boldsymbol{B}) + \boldsymbol{v} \cdot (\nabla_x \boldsymbol{B}) - \boldsymbol{B} \cdot (\nabla_x \boldsymbol{v}), \tag{3.41}$$

to yield

$$\begin{aligned}
\frac{d}{dt} \int_{A_\Omega} \boldsymbol{B} \cdot \boldsymbol{n}\, dA &= \int_{A_\Omega} \left(\frac{d}{dt} \boldsymbol{B} + \boldsymbol{B} \nabla_x \cdot \boldsymbol{v} - \boldsymbol{B} \cdot \nabla_x \boldsymbol{v} \right) \cdot \boldsymbol{n}\, dA \\
&= \int_{A_\Omega} \left(\frac{\partial}{\partial t} \boldsymbol{B} + \underbrace{(\nabla_x \cdot \boldsymbol{B})}_{=0} \boldsymbol{v} + \nabla_x \times (\boldsymbol{B} \times \boldsymbol{v}) \right) \cdot \boldsymbol{n}\, dA \\
&= \int_{A_\Omega} \frac{\partial \boldsymbol{B}}{\partial t} \cdot \boldsymbol{n}\, dA + \int_{C_\Omega} (\boldsymbol{B} \times \boldsymbol{v}) \cdot d\boldsymbol{L}.
\end{aligned} \tag{3.42}$$

This result is sometimes referred to as "Lenz's" law, and is closely related to Zorawski's criterion. For more details, see Chandrasekharaiah and Debnath [10], Malvern [64], or Jackson [46].

Remark 3: One can interpret the EMF as having two contributions:

1. Due to changes in the magnetic field with the medium fixed ($\frac{\partial \boldsymbol{B}}{\partial t}$) and
2. Due to changes in the configuration of the medium and the magnetic field (due to the Lorentz contribution ($\boldsymbol{v} \times \boldsymbol{B}$)).

Explicitly, the contributions are

- Due to changes in the magnetic field: $\int_{C_\Omega} \frac{\partial \boldsymbol{B}}{\partial t} \cdot \boldsymbol{n}\, dA$,
- Due to shape or configuration changes of the medium, which can be interpreted as arising from the magnetic-medium-movement contribution to the Lorentz force, per unit charge: $\frac{\boldsymbol{\Psi}^m}{q} = \boldsymbol{v} \times \boldsymbol{B}$, leading to: $\int_{C_\Omega} (\boldsymbol{B} \times \boldsymbol{v}) \cdot d\boldsymbol{L}$.

Equating the negation of these contributions to the electric field contour yields

$$\frac{d}{dt}\int_{C\Omega} \boldsymbol{B} \cdot d\boldsymbol{L} = \underbrace{\int_{A\Omega} \frac{\partial \boldsymbol{B}}{\partial t} \cdot \boldsymbol{n}\, dA}_{\text{field changes}} + \underbrace{\int_{C\Omega} \boldsymbol{B} \times \boldsymbol{v} \cdot d\boldsymbol{L}}_{\text{configuration changes}} . \tag{3.43}$$

3.8 Constitutive Relations

The following linear constitutive laws are commonly employed:

$$\boldsymbol{D} = \underbrace{\boldsymbol{\epsilon} \cdot \boldsymbol{E}}_{\text{anisotropic}} = \underbrace{\epsilon \boldsymbol{E}}_{\text{isotropic}} \tag{3.44}$$

and

$$\boldsymbol{B} = \underbrace{\boldsymbol{\mu} \cdot \boldsymbol{H}}_{\text{anisotropic}} = \underbrace{\mu \boldsymbol{H}}_{\text{isotropic}} , \tag{3.45}$$

and (under the assumption that $\boldsymbol{v} = \boldsymbol{0}$)

$$\boldsymbol{J} = \underbrace{\boldsymbol{\sigma} \cdot (\boldsymbol{E} + \boldsymbol{v} \times \boldsymbol{B})}_{\text{anisotropic}} = \underbrace{\sigma(\boldsymbol{E} + \boldsymbol{v} \times \boldsymbol{B})}_{\text{isotropic}} = \underbrace{\sigma \boldsymbol{E}}_{if\ \boldsymbol{v}=\boldsymbol{0}} \tag{3.46}$$

where, here, σ is the conductivity and

$$\boldsymbol{M}_c = \underbrace{\hat{\boldsymbol{\sigma}} \cdot \boldsymbol{H}}_{\text{anisotropic}} = \underbrace{\hat{\sigma} \boldsymbol{H}}_{\text{isotropic}} , \tag{3.47}$$

where, here, $\hat{\sigma}$ is the equivalent (phenomenological) magnetic "conductivity."
Such (linear, isotropic) responses stem from the decomposition[3]

$$\boldsymbol{D} = \epsilon_o(1 + \chi_E)\boldsymbol{E}, \tag{3.48}$$

where χ_E is the electric susceptibility that typically ranges from $0 \le \chi_E \le 10^4$. Similarly, for the magnetic properties,

$$\boldsymbol{B} = \mu_o(1 + \chi_H)\boldsymbol{H}, \tag{3.49}$$

[3] These relations can easily be extended to anisotropic cases as well.

where χ_H is the magnetic susceptibility that typically ranges from $0 \leq \chi_H \leq 10^4$. Such relations can be generalized further to extend into the nonlinear regime which accounts for polarization and magnetization effects

$$\boldsymbol{D} = \mathcal{E}(\boldsymbol{E}, \boldsymbol{\epsilon}, etc.) \tag{3.50}$$

and

$$\boldsymbol{B} = \mathcal{M}(\boldsymbol{H}, \boldsymbol{\mu}, etc.). \tag{3.51}$$

Polarization describes the separation of positive and negative charges at the atomic and molecular scales within solids. Dielectric materials become polarized in an electric field, resulting in the electric field flux density, $\boldsymbol{D}$, becoming larger than in the same applied electric field in free-space ($\epsilon \geq \epsilon_o$).

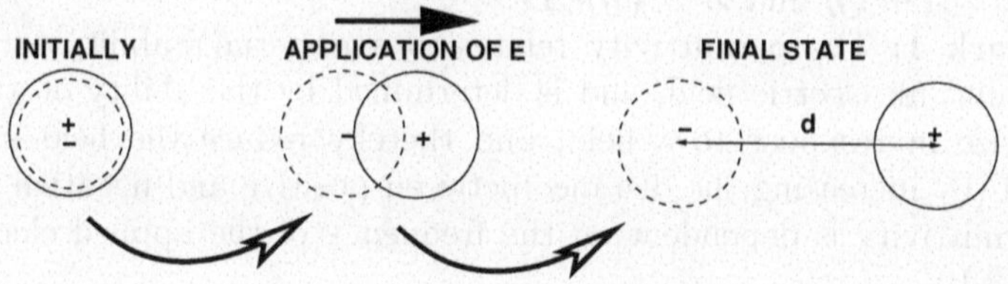

Fig. 3.9. A sequence of events resulting in a polarization of charges.

A simple model of polarization is to treat the initial positive and negative charges within an atom as being "bound" overlapping regions, which upon the introduction of an electric field $\boldsymbol{E}$, become successively separated (Figure 3.9), with the positive charge moving in the direction of the field and the negative charge moving in the opposite direction. The displacement can be described by an electric "dipole" moment $\boldsymbol{p}$ (see Figure 3.9).

$$\boldsymbol{p} = \mathcal{P}\boldsymbol{d}. \tag{3.52}$$

For most materials, the charged regions will return to their original state if the field is removed.If we consider a small region δV of polarized dielectric material, containing N dipole moments $\boldsymbol{p}$, the polarization $\boldsymbol{P}$ is described as the dipole moment per unit volume:

$$\boldsymbol{P} = lim_{\delta V \to 0} \frac{N\boldsymbol{p}}{\delta V}, \tag{3.53}$$

which is normally introduced into constitutive relations (in the isotropic case) through

$$\boldsymbol{D} = \epsilon_o \boldsymbol{E} + \boldsymbol{P}, \tag{3.54}$$

where $\boldsymbol{P} = \chi_E \epsilon_o \boldsymbol{E}$, and therefore

$$\boldsymbol{D} = \epsilon_o \boldsymbol{E} + \boldsymbol{P} = \epsilon_o (1 + \chi_E) \boldsymbol{E} = \epsilon_o \epsilon_r \boldsymbol{E} = \epsilon \boldsymbol{E}, \tag{3.55}$$

where $\epsilon_r = 1 + \chi_E$. The magnetic permeability is described in a similar manner

$$\boldsymbol{B} = \mu_o \boldsymbol{H} + \boldsymbol{Z} = \mu_o (1 + \chi_H) \boldsymbol{H} = \mu_o \mu_r \boldsymbol{H} = \mu \boldsymbol{H}, \tag{3.56}$$

where $\mu_r = 1 + \chi_H$ and $\boldsymbol{Z} = \chi_H \mu_o \boldsymbol{H}$.

Remark 1: The permittivity relates to a material's ability to transmit or "permit" an electric field, and is determined by the ability of a material to polarize in response to a field, and thereby reduce the field inside the material (by increasing the distance between positive and negative charges). The permittivity is dependent on the frequency of the applied electric field (Figure 3.10):

1. In the low-frequency regime $(0 - 10^9)$Hz, ionic effects control the polarization/permittivity,
2. In the medium-frequency regime $(10^9 - 10^{12})$Hz, dipole effect and bond relaxation control the polarization/permittivity,
3. In the high-frequency regime $(10^{12} - 10^{15})$Hz, atomic resonances control the polarization/permittivity and
4. In the ultrahigh-frequency regime $(>> 10^{15})$Hz, electron effects (resonances) control the polarization/permittivity.

However, over the range of frequencies of interest here, $(0 - 10^9)$Hz, ϵ is more or less frequency-independent (Figure 3.10).[4] *We will concentrate primarily on linear constitutive laws in this work.* For an in-depth discussion of nonlinear electromagnetic responses, see Jackson [46].

Remark 2: Generally, one should interpret all quantities on the continuum scale as being the homogenized (averaged) effects of multiple charges within atoms, molecules, etc. One could take terms beyond $\boldsymbol{P}$ and $\boldsymbol{Z}$ (tripoles,

[4] Media in which ϵ depends on the frequency are termed "dispersive."

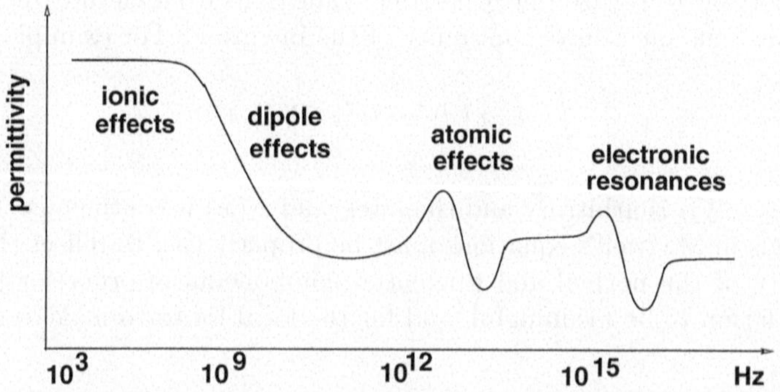

Fig. 3.10. Variation of the permittivity over frequency of electric fields.

quadrapoles, etc.). The preceding discussion is a gross simplification and we again refer the reader to Jackson [46] for a thorough account.

Remark 3: Some materials, so-called ferromagnets, are inherently magnetic by nature. They are composed of a large number of "domains" which are not always aligned, and act against one another to reduce the net magnetic field. However, upon application of an external magnetic field, the domains tend to line up so that the sum of the fields from the domains is higher than the applied magnetic field alone.

Remark 4: In the case of the electrical conductivity, one may write

$$\sigma = nq(m_e + m_h) \tag{3.57}$$

where $n \propto e^{-E_s/2k\theta}$ is the number of electrons, E_s is the band gap energy (needed for an electron to leave its valence shell and to become available for conduction), k is the Boltzmann constant, θ is the absolute temperature, q is charge of an individual electron, m_e is the mobility of the electrons and m_h is the mobility of the holes. Thus, we have

$$\sigma = Ce^{-E_s/2k\theta}q(m_e + m_h) = \sigma_{\theta_o}e^{K_\sigma(\theta-\theta_o)}, \tag{3.58}$$

where C is a preexponential material constant and K_σ is a thermal growth constant.

3.9 Localization: Point-Forms of Maxwell's Equations

It is important to realize that when one transitions from an integral to pointwise forms, one must have continuity of the integrand. For example,

$$\int_{V_\omega} F(\boldsymbol{x})\, dV = 0 \Rightarrow F(\boldsymbol{x}) = 0, \tag{3.59}$$

only if $V_\omega \subset V_\Omega$ is arbitrary and the integrand $F(\boldsymbol{x})$ is continuous. Thus, the integrands in Maxwell's equations must be properly cast to reflect the proper continuity of the normal and tangential components in order for pointwise differentiation to be meaningful, and for the local forms to make sense.

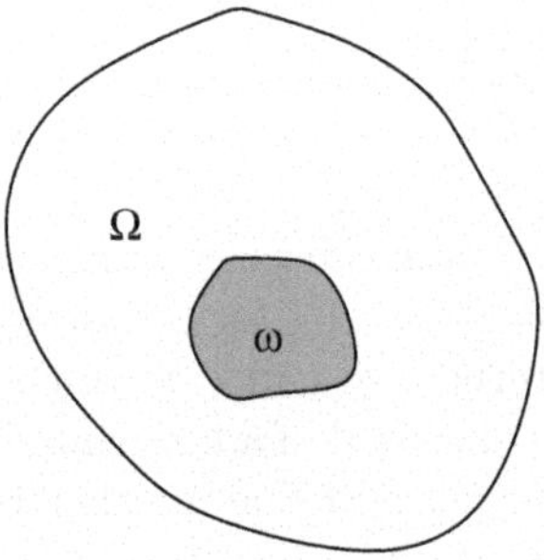

Fig. 3.11. An arbitrary portion of continua.

Isolating an arbitrary portion of the domain $V_\omega \subset V_\Omega$, in integral form, we have

$$\int_{C_\omega} \boldsymbol{E} \cdot d\boldsymbol{L} = -\int_{A_\omega} (\boldsymbol{M}^{tot} + \frac{\partial \boldsymbol{B}}{\partial t}) \cdot \boldsymbol{n}\, dA \tag{3.60}$$

and

$$\int_{C_\omega} \boldsymbol{H} \cdot d\boldsymbol{L} = \int_{A_\omega} (\boldsymbol{J}^{tot} + \frac{\partial \boldsymbol{D}}{\partial t}) \cdot \boldsymbol{n}\, dA, \tag{3.61}$$

and

$$\int_{A_\omega} \boldsymbol{D} \cdot \boldsymbol{n} \, dA = \int_{V_\omega} \mathcal{P} \, dV, \tag{3.62}$$

and

$$\int_{A_\omega} \boldsymbol{B} \cdot \boldsymbol{n} \, dA = 0. \tag{3.63}$$

We must employ two tools, that is, the Divergence theorem ($\mathcal{F}$ is an arbitrary vector)

$$\int_{A_\omega} \mathcal{F} \cdot \boldsymbol{n} \, dA = \int_{V_\omega} \nabla_x \cdot \mathcal{F} \, dV \tag{3.64}$$

and (2) Stoke's theorem

$$\int_{C_\omega} \mathcal{F} \cdot d\boldsymbol{L} = \int_{A_\omega} (\nabla_x \times \mathcal{F}) \cdot \boldsymbol{n} \, dA. \tag{3.65}$$

Thus, for Equation 3.60, applying Stoke's theorem and since ω is arbitrary,

$$\int_{C_\omega} \boldsymbol{H} \cdot d\boldsymbol{L} = \int_{A_\omega} (\nabla_x \times \boldsymbol{E}) \cdot \boldsymbol{n} \, dA = -\int_{A_\omega} (\boldsymbol{M}^{tot} + \frac{\partial \boldsymbol{B}}{\partial t}) \cdot \boldsymbol{n} \, dA \Rightarrow \nabla_x \times \boldsymbol{E} = -\boldsymbol{M}^{tot} - \frac{\partial \boldsymbol{B}}{\partial t}. \tag{3.66}$$

Applying Stoke's theorem to Equation 3.66 and thus, since ω is arbitrary,

$$\int_{C_\omega} \boldsymbol{H} \cdot d\boldsymbol{L} = \int_{A_\omega} (\nabla_x \times \boldsymbol{H}) \cdot \boldsymbol{n} \, dA = \int_{A_\omega} (\boldsymbol{J}^{tot} + \frac{\partial \boldsymbol{D}}{\partial t}) \cdot \boldsymbol{n} \, dA \Rightarrow \nabla_x \times \boldsymbol{H} = \boldsymbol{J}^{tot} + \frac{\partial \boldsymbol{D}}{\partial t}. \tag{3.67}$$

For Equation 3.62,

$$\int_{A_\omega} \boldsymbol{D} \cdot \boldsymbol{n} \, dA = \int_{V_\omega} \nabla_x \cdot \boldsymbol{D} \, dV = \int_{V_\omega} \mathcal{P} \, dV \Rightarrow \nabla_x \cdot \boldsymbol{D} = \mathcal{P}. \tag{3.68}$$

Finally,

$$\int_{A_\omega} \boldsymbol{B} \cdot \boldsymbol{n} \, dA = \int_{V_\omega} \nabla_x \cdot \boldsymbol{B} \, dV = 0 \Rightarrow \nabla_x \cdot \boldsymbol{B} = 0. \tag{3.69}$$

Remark: As mentioned earlier, we will introduce phenomenological terms that produce an effect that leads to $\nabla_x \cdot \boldsymbol{B} \neq 0$.

3.10 Remark on Conservation Laws

Consider

$$\nabla_x \times \boldsymbol{H} = \boldsymbol{J}^{tot} + \frac{\partial \boldsymbol{D}}{\partial t}, \tag{3.70}$$

and taking the divergence

$$\nabla_x \cdot (\nabla_x \times \boldsymbol{H}) = 0 = \nabla_x \cdot (\boldsymbol{J}^{tot} + \frac{\partial \boldsymbol{D}}{\partial t}) = 0 \Rightarrow \nabla_x \cdot \boldsymbol{D} = \mathcal{P}, \tag{3.71}$$

provided that

$$\nabla_x \cdot \boldsymbol{J}^{tot} + \frac{\partial \mathcal{P}}{\partial t} = 0, \tag{3.72}$$

or that a charge conservation is obeyed. Now, consider

$$\nabla_x \times \boldsymbol{E} = -(\boldsymbol{M}^{tot} + \frac{\partial \boldsymbol{B}}{\partial t}), \tag{3.73}$$

and taking the divergence

$$\nabla_x \cdot (\nabla_x \times \boldsymbol{E}) = 0 = -\nabla_x \cdot (\boldsymbol{M}^{tot} + \frac{\partial \boldsymbol{B}}{\partial t}) = 0 \Rightarrow \nabla_x \cdot \boldsymbol{B} = \mathcal{F}, \tag{3.74}$$

provided that

$$\nabla_x \cdot \boldsymbol{M}^{tot} = -\frac{\partial \mathcal{F}}{\partial t}, \tag{3.75}$$

or that a *phenomenological* "magnetic" charge conservation is obeyed. *The conclusion is that solving the curl equations implies solving the divergence equations, automatically, if the charge conservation laws are obeyed.*

Observation: The current/charge conservation law in Equation 3.72 is somewhat different than what one might expect, for example, for fluid flow, where a total derivative ($\frac{d}{dt}$) would be in place of a partial derivative ($\frac{\partial}{\partial t}$).

3.11 Explicit Representations

One can decompose the terms $\boldsymbol{M}$ and $\boldsymbol{J}$ into conductive and sources

$$\boldsymbol{J}^{tot} = \sigma(\boldsymbol{E} + \boldsymbol{v} \times \boldsymbol{B}) + \boldsymbol{J}^{ext}, \tag{3.76}$$

and

$$\boldsymbol{M}^{tot} = \hat{\sigma}\boldsymbol{H} + \boldsymbol{M}^{ext}. \tag{3.77}$$

This leads to

$$\nabla_x \times \boldsymbol{E} = -(\boldsymbol{M}^{ext} + \hat{\sigma}\boldsymbol{H} + \frac{\partial \boldsymbol{B}}{\partial t}), \tag{3.78}$$

and

$$\nabla_x \times \boldsymbol{H} = \sigma(\boldsymbol{E} + \boldsymbol{v} \times \boldsymbol{B}) + \boldsymbol{J}^{ext} + \frac{\partial \boldsymbol{D}}{\partial t}. \tag{3.79}$$

3.12 Interface Conditions

The interface conditions at the boundaries/interfaces can be derived from the field equations themselves by considering special contour and area integrals.

3.12.1 The Electric Field Intensity

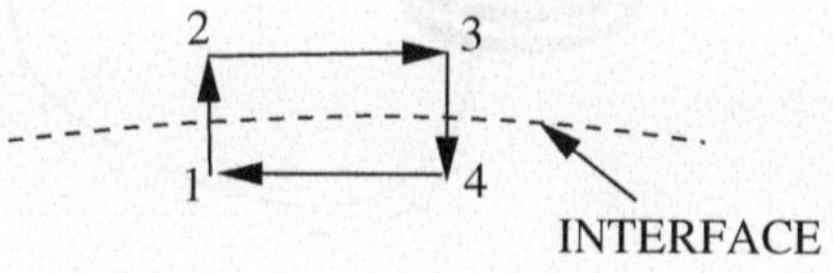

Fig. 3.12. An interface circuit.

For the electric field, let us specialize Ampere's law

$$\int_C \boldsymbol{E} \cdot d\boldsymbol{L} = -\int_A (\boldsymbol{M}^{tot} + \frac{\partial \boldsymbol{B}}{\partial t}) \cdot \boldsymbol{n}\, dA \tag{3.80}$$

with the specific contour in Figure 3.12, that is,

$$\int_C \boldsymbol{E} \cdot d\boldsymbol{L} = \int_1^2 \boldsymbol{E} \cdot d\boldsymbol{s} + \int_2^3 \boldsymbol{E} \cdot d\boldsymbol{s} + \int_3^4 \boldsymbol{E} \cdot d\boldsymbol{s} + \int_4^1 \boldsymbol{E} \cdot d\boldsymbol{s} = -\int_{A_\Omega} (\boldsymbol{M}^{tot} + \frac{\partial \boldsymbol{B}}{\partial t}) \cdot \boldsymbol{n}\, dA, \tag{3.81}$$

where $\boldsymbol{ds} = \boldsymbol{t}\,dA$, where $\boldsymbol{t}$ is the tangent vector. Now, collapse $1 \rightarrow 2$ and $3 \rightarrow 4$, forcing the integral on the right to vanish since there are no concentrated magnetic fields, thus leading to

$$\int_2^3 \boldsymbol{E}\cdot\boldsymbol{ds}+\int_4^1 \boldsymbol{E}\cdot\boldsymbol{ds} = 0 \Rightarrow \boldsymbol{t}_{12}\cdot(\boldsymbol{E}_2-\boldsymbol{E}_1) = 0 \Rightarrow \boldsymbol{n}_{12}\times(\boldsymbol{E}_2-\boldsymbol{E}_1) = \boldsymbol{0} \quad (3.82)$$

which indicates that the tangential component of the electric field must be continuous.

3.12.2 Electric Flux Density

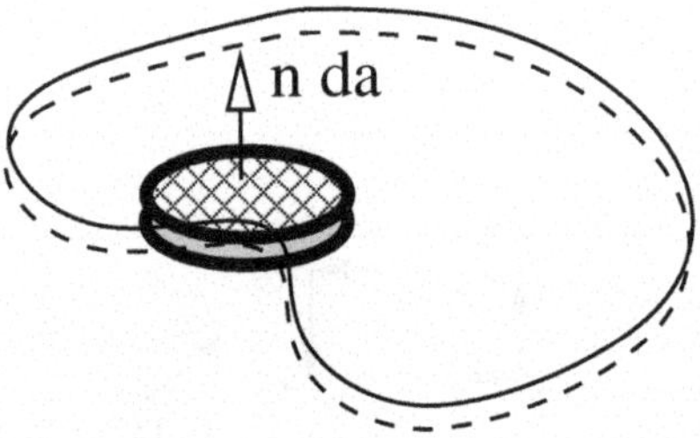

Fig. 3.13. An interface.

For the electric flux density, using Gauss's saw (Figure 3.13), one isolates a portion of the boundary (Figure 3.13)

$$\int_A \boldsymbol{D}\cdot\boldsymbol{n}\,dA = \int_{top} \boldsymbol{D}\cdot\boldsymbol{n}\,dA+\int_{bottom} \boldsymbol{D}\cdot\boldsymbol{n}\,dA+\int_{sides} \boldsymbol{D}\cdot\boldsymbol{n}\,dA = \int_{interface} \mathcal{P}\,dA \quad (3.83)$$

where $\mathcal{P}$ is a concentrated surface charge between media. Now, collapse the cylinder

$$\int_{top} \boldsymbol{D}\cdot\boldsymbol{n}\,dA + \int_{bottom} \boldsymbol{D}\cdot\boldsymbol{n}\,dA = \int_{interface} \mathcal{P}\,dA \Rightarrow (\boldsymbol{D}_2 - \boldsymbol{D}_1)\cdot\boldsymbol{n}_{12} = 0, \quad (3.84)$$

which indicates that the normal component of the electric field flux intensity must be continuous, *provided that $\mathcal{P}$ is not a localized (concentrated) sheet of charges.*

3.12.3 The Magnetic Field Intensity

For the magnetic field, let us specialize the contour integral

$$\int_C \boldsymbol{H} \cdot d\boldsymbol{L} = \int_A (\boldsymbol{J} + \frac{\partial \boldsymbol{D}}{\partial t}) \cdot \boldsymbol{n}\, dA. \tag{3.85}$$

with the specific contour in Figure 3.12.

$$\int_C \boldsymbol{H} \cdot d\boldsymbol{L} = \int_1^2 \boldsymbol{H} \cdot d\boldsymbol{s} + \int_2^3 \boldsymbol{H} \cdot d\boldsymbol{s} + \int_3^4 \boldsymbol{H} \cdot d\boldsymbol{s} + \int_4^1 \boldsymbol{H} \cdot d\boldsymbol{s} = \int_{A_\Omega} (\boldsymbol{J} + \frac{\partial \boldsymbol{D}}{\partial t}) \cdot \boldsymbol{n}\, dA, \tag{3.86}$$

where $d\boldsymbol{s} = \boldsymbol{t}\, dA$. Now, collapse $1 \to 2$ and $3 \to 4$. The integral on the right does not necessarily vanish since there can be concentrated electric currents, leading to

$$\int_2^3 \boldsymbol{H} \cdot d\boldsymbol{s} + \int_4^1 \boldsymbol{H} \cdot d\boldsymbol{s} = \int_{A_\Omega} \boldsymbol{j}_s\, dA \Rightarrow \boldsymbol{t}_{12} \cdot (\boldsymbol{H}_2 - \boldsymbol{H}_1) = j_s \Rightarrow \boldsymbol{n}_{12} \times (\boldsymbol{H}_2 - \boldsymbol{H}_1) = \boldsymbol{j}_s \tag{3.87}$$

indicating that the tangential component of the electric field must be continuous if there are no concentrated electric currents. Here, $\mathbf{j}_s = j_s \boldsymbol{t}$ is the surface current density between surfaces on the tangent interface plane.

3.12.4 Magnetic Flux

For the magnetic flux, using Gauss's law (Figure 3.13), one isolates a portion of the boundary (Figure 3.13)

$$\int_A \boldsymbol{B} \cdot \boldsymbol{n}\, dA = \int_{top} \boldsymbol{B} \cdot \boldsymbol{n}\, dA + \int_{bottom} \boldsymbol{B} \cdot \boldsymbol{n}\, dA + \int_{sides} \boldsymbol{B} \cdot \boldsymbol{n}\, dA = 0, \tag{3.88}$$

where the right-hand side is zero because there are no monopoles (concentrated magnetic "charges") between the media. Now, collapse the cylinder

$$\int_{top} \boldsymbol{B} \cdot \boldsymbol{n}\, dA + \int_{bottom} \boldsymbol{B} \cdot \boldsymbol{n}\, dA = 0 \Rightarrow (\boldsymbol{B}_2 - \boldsymbol{B}_1) \cdot \boldsymbol{n}_{12} = 0, \tag{3.89}$$

indicating that the normal component of the magnetic flux must be continuous.

3.12.5 Summary

Thus, when there is no concentrated charge or current on an interface or boundary, the following must hold:

- continuity of the tangential components of the electric field at an interface: $\boldsymbol{t}_{12} \cdot (\boldsymbol{E}_1 - \boldsymbol{E}_2) = 0$ or $\boldsymbol{n}_{12} \times (\boldsymbol{E}_1 - \boldsymbol{E}_2) = \boldsymbol{0}$,
- continuity of the tangential components of the magnetic field at an interface: $\boldsymbol{t}_{12} \cdot (\boldsymbol{H}_1 - \boldsymbol{H}_2) = 0$ or $\boldsymbol{n}_{12} \times (\boldsymbol{H}_1 - \boldsymbol{H}_2) = \boldsymbol{0}$,
- continuity of the electric flux normal to the interface: $\boldsymbol{n}_{12} \cdot (\boldsymbol{D}_1 - \boldsymbol{D}_2) = 0$,
- continuity of the magnetic flux normal to the interface: $\boldsymbol{n}_{12} \cdot (\boldsymbol{B}_1 - \boldsymbol{B}_2) = 0$.

3.13 Physical Interpretation of Operators

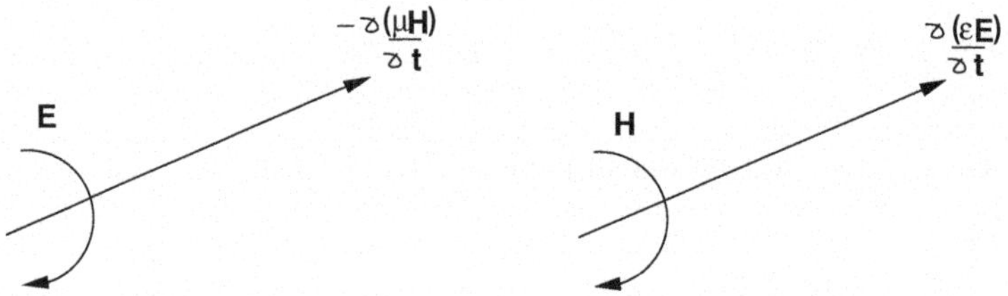

Fig. 3.14. The axis of rotation interpretation.

In order to physically interpret Maxwell's equations, we investigate operators that appear therein.

Curl: The curl of a vector field $\boldsymbol{A}$, $\nabla_x \times \boldsymbol{A}$, indicates the vector field's rate of rotation (Figure 3.14). In particular, the axis of rotation (right-hand rule) is $\nabla_x \times \boldsymbol{A}$ and the magnitude of rotation is $||\nabla_x \times \boldsymbol{A}||$. If $\nabla_x \times \boldsymbol{A} = \boldsymbol{0}$, then $\boldsymbol{A}$ is said to be "irrotational."

Divergence: The divergence is an operator that measures the magnitude of a vector field's source or sink at a given point, $\nabla_x \cdot \boldsymbol{A}$. For example, for a vector field representing the velocity of air expanding as it is heated, the divergence of velocity field would have positive values if expanding and negative values if contracting (cooling). If $\nabla_x \cdot \boldsymbol{A} = 0$, $\boldsymbol{A}$ is said to be "solenoidal." The gradient of a scalar is a vector field which points in the direction of the greatest rate of increase and whose magnitude is the greatest rate of change. Thus, for an electromagnetic medium that is nonconducting and source-free, we have

$$\nabla_x \times \boldsymbol{E} = -\frac{\partial(\mu \boldsymbol{H})}{\partial t}, \tag{3.90}$$

which is the axis of rotation of $\boldsymbol{E}$, and $||\frac{\partial(\mu \boldsymbol{H})}{\partial t}||$ is the magnitude of that rotation. *The time-rate of change of $\boldsymbol{H}$ dictates the rate of rotation of $\boldsymbol{E}$.* Conversely,

$$\nabla_x \times \boldsymbol{H} = \frac{\partial(\epsilon \boldsymbol{E})}{\partial t} \tag{3.91}$$

is the axis of rotation of $\boldsymbol{H}$ and $||\frac{\partial(\epsilon \boldsymbol{E})}{\partial t}||$ is the magnitude. The time-rate of change of $\boldsymbol{E}$ dictates the rate of rotation of $\boldsymbol{H}$. In each case, one serves as the axis of rotation for the other. Also, clearly, if there are no external charges, $\nabla_x \cdot \boldsymbol{D} = 0$ and $\nabla_x \cdot \boldsymbol{B} = 0$, then $\boldsymbol{D}$ and $\boldsymbol{B}$ are solenoidal.

3.14 "Hidden" Maxwellian Waves

Consider a homogeneous electromagnetic medium with no conduction or source terms. Applying the vector identity to Maxwell's equations [5]

$$\nabla_x \times (\nabla_x \times \boldsymbol{A}) = \nabla_x(\nabla_x \cdot \boldsymbol{A}) - \nabla_x^2 \boldsymbol{A}, \tag{3.92}$$

we obtain

$$\nabla_x^2 \boldsymbol{E} = \mu \frac{\partial}{\partial t}(\nabla_x \times \boldsymbol{H}) = \mu\epsilon \frac{\partial^2 \boldsymbol{E}}{\partial t^2} \tag{3.93}$$

and

[5] Note, using the vector identity $\nabla_x \cdot (\nabla_x \times \boldsymbol{A}) = 0$, we have $\nabla_x \cdot \boldsymbol{H} = 0$ and $\nabla_x \cdot \boldsymbol{E} = 0$.

$$\nabla_x^2 \boldsymbol{H} = -\epsilon \frac{\partial}{\partial t}(\nabla_x \times \boldsymbol{E}) = \mu\epsilon \frac{\partial^2 \boldsymbol{H}}{\partial t^2}. \tag{3.94}$$

When there is no external source charge, no conductivity and no external current, then one can show that $\boldsymbol{E}$ is both in phase with and orthogonal to $\boldsymbol{H}$. Maxwell's equations take the form

$$\nabla_x^2 \boldsymbol{E} = \frac{1}{c^2}\frac{\partial^2 \boldsymbol{E}}{\partial t^2} \tag{3.95}$$

and

$$\nabla_x^2 \boldsymbol{H} = \frac{1}{c^2}\frac{\partial^2 \boldsymbol{H}}{\partial t^2}, \tag{3.96}$$

where c is the wave speed in the medium. These describe transverse waves; i.e., they cause "disturbances" in the medium in a perpendicular direction to the direction of propagation. For example, if a wave moves in the x-direction, the disturbances are in the y-z plane (Figure 3.15). We make the following observations:

1. $\nabla_x \cdot \boldsymbol{E} = 0$ and $\nabla_x \cdot \boldsymbol{H} = 0$ imply that they are transverse waves.
2. $\nabla_x \times \boldsymbol{E} = -\mu_o \frac{\partial \boldsymbol{H}}{\partial t}$ and $\nabla_x \times \boldsymbol{H} = \epsilon_o \frac{\partial \boldsymbol{E}}{\partial t}$ imply that the wave of the electric field $\boldsymbol{E}$ is in phase with $\boldsymbol{H}$ and perpendicular to it.

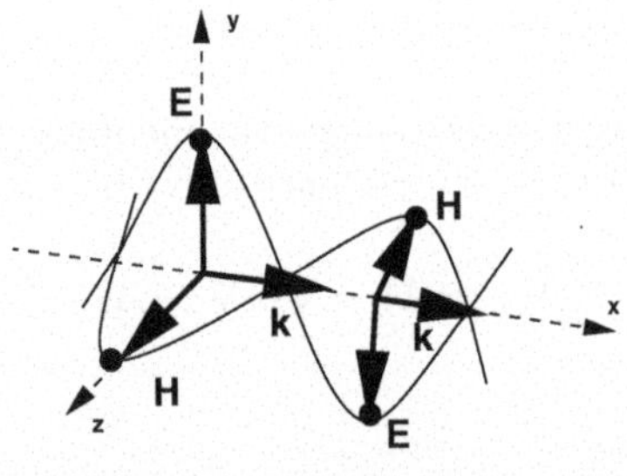

Fig. 3.15. The transverse interpretation.

In order to see the implications of (1) and (2), consider a generic vector field for $\boldsymbol{E}$

$$\boldsymbol{E} = \boldsymbol{E}_o f(\boldsymbol{k}, \boldsymbol{r}, c, t), \tag{3.97}$$

where $\boldsymbol{E}_o$ is the constant amplitude. Here, f is a twice differentiable function, and $\boldsymbol{k}$ is the unit vector in the direction of propagation and $\boldsymbol{r}$ is the position vector. f satisfies the wave equation

$$\nabla_x^2 f = \frac{1}{c^2}\frac{\partial^2 f}{\partial t^2}. \tag{3.98}$$

This form will satisfy the wave equation. We have the following further observations:

- $\nabla_x \cdot \boldsymbol{E} = \nabla_x \cdot (\boldsymbol{E}_o f\underbrace{(\boldsymbol{k}\cdot\boldsymbol{r} - ct)}_{\phi}) = (\boldsymbol{E}_o \cdot \boldsymbol{k})\frac{\partial f}{\partial \phi} \Rightarrow \boldsymbol{E}$ and $\boldsymbol{k}$ are orthogonal,
- $\nabla_x \times \boldsymbol{E} = (\boldsymbol{k}\times\boldsymbol{E})\frac{\partial f}{\partial \phi} = -\mu\frac{\partial \boldsymbol{H}}{\partial t} = \mu c\boldsymbol{H}\frac{\partial f}{\partial \phi} \Rightarrow \boldsymbol{k}\times\boldsymbol{E} = \mu c\boldsymbol{H} \Rightarrow \boldsymbol{E}$ and $\boldsymbol{H}$ are perpendicular,
- $\nabla_x \cdot \mu\boldsymbol{H} = \mu\boldsymbol{k}\cdot\boldsymbol{H}\frac{\partial f}{\partial \phi} = 0 \Rightarrow \boldsymbol{k}\cdot\boldsymbol{H} = 0 \Rightarrow \boldsymbol{H}$ and $\boldsymbol{k}$ are perpendicular,
- $\nabla_x \times \boldsymbol{H} = (\boldsymbol{k}\times\boldsymbol{H})\frac{\partial f}{\partial \phi} = \epsilon\frac{\partial \boldsymbol{E}}{\partial t} = -\epsilon c\boldsymbol{E}\frac{\partial f}{\partial \phi} \Rightarrow \boldsymbol{k}\times\boldsymbol{H} = -\epsilon c\boldsymbol{E} \Rightarrow \boldsymbol{E}$ and $\boldsymbol{H}$ are perpendicular (as before).

Thus, $||\boldsymbol{E}_o|| = |\mu c|||\boldsymbol{H}_o||$ and $||\boldsymbol{H}_o|| = |\epsilon c|||\boldsymbol{E}_o|| = \frac{1}{|\mu c|}||\boldsymbol{E}_o|| \Rightarrow$ and $c = \frac{1}{\sqrt{\epsilon\mu}}$. Also, this implies that $||\boldsymbol{B}_o|| = \frac{1}{|c|}||\boldsymbol{E}_o||$.

Remark 1: Also, in electro- and magnetostatics, when $\frac{\partial \boldsymbol{D}}{\partial t} = 0$ and $\frac{\partial \boldsymbol{B}}{\partial t} = 0$, the two fields are irrotational (curl-free), $\nabla_x \times \boldsymbol{E} = \boldsymbol{0}$ and $\nabla_x \times \boldsymbol{H} = \boldsymbol{0}$. If the loading is gradual enough, wave effects will be very mild or nonexistent. Wave-like phenomena will not be explored in this monograph.

Remark 2: Although we will not analyze permanent magnets in this monograph, we provide a few comments. The atomic origin of permanent magnetism arises from the fact that electrons, which can be thought of as "little (dipole) magnets," have aligned spins in a permanent magnet, as opposed to randomly oriented spins in an "ordinary" material. A simple approach to account for permanent magnetism is through the constitutive law

$$\boldsymbol{B} = \boldsymbol{\mu}\cdot(\boldsymbol{H} + \boldsymbol{H}^{Perm}), \tag{3.99}$$

where $\boldsymbol{H}^{perm}$ is the permanent magnetic field. This constitutive law is then used in Maxwell's equations, specifically

$$\nabla_x \times (\boldsymbol{H} + \boldsymbol{H}^{perm}) = \boldsymbol{J}. \tag{3.100}$$

4

Classical Linear Constitutive Behavior

We now consider idealized linear material behavior.

4.1 Linear Constitutive Equations

The starting point to develop a constitutive theory is to assume that an energy function per unit volume exists, a nonnegative function denoted W. A linear constitutive relation can be derived from

$$\boldsymbol{D} = \frac{\partial W}{\partial \boldsymbol{E}} \tag{4.1}$$

and

$$W \approx c_0 + \mathbf{c}_1 \cdot \boldsymbol{E} + \frac{1}{2}\boldsymbol{E} \cdot \boldsymbol{\epsilon} \cdot \boldsymbol{E} + ... \tag{4.2}$$

which implies

$$\boldsymbol{D} \approx \mathbf{c}_1 + \boldsymbol{\epsilon} \cdot \boldsymbol{E} + ... \tag{4.3}$$

We are free to set $c_0 = 0$ (it is arbitrary) to have zero energy at zero $\boldsymbol{E}$, and furthermore, we assume that no electric field fluxes exist in the reference state ($\mathbf{c}_1 = \mathbf{0}$). Thus, we obtain a general anisotropic relation between the electric flux and the electric field:

$$\boldsymbol{D} = \boldsymbol{\epsilon} \cdot \boldsymbol{E}. \tag{4.4}$$

This is a linear (tensorial) relation. A general material has nine independent constants since it is a second order tensor relating three components of electric flux to electric field. This is easily seen from the matrix representation of $\boldsymbol{\epsilon}$

T.I. Zohdi: Electromagnetic Properties of Multiphase Dielectrics, LNACM 64, pp. 39–43.
springerlink.com

$$\underbrace{\begin{Bmatrix} D_1 \\ D_2 \\ D_3 \end{Bmatrix}}_{\stackrel{\text{def}}{=}\{\boldsymbol{D}\}} = \underbrace{\begin{bmatrix} \epsilon_{11} & \epsilon_{12} & \epsilon_{13} \\ \epsilon_{21} & \epsilon_{22} & \epsilon_{23} \\ \epsilon_{31} & \epsilon_{32} & \epsilon_{33} \end{bmatrix}}_{\stackrel{\text{def}}{=}[\boldsymbol{\epsilon}]} \underbrace{\begin{Bmatrix} E_1 \\ E_2 \\ E_3 \end{Bmatrix}}_{\stackrel{\text{def}}{=}\{\boldsymbol{E}\}}. \tag{4.5}$$

The symbol $[\cdot]$ is used to indicate standard matrix notation equivalent to a tensor form. At this point, based on many factors that depend on the material microstructure, it can be shown that the components of $\boldsymbol{\epsilon}$ may be written in terms of anywhere between nine and one independent parameters. We explore such concepts further via the ideas of material symmetry.

Remark: The existence of a strictly positive stored energy function implies that the linear tensor must have positive eigenvalues at every point in the body. Typically, different materials are classified according to the number of independent constants in $\boldsymbol{\epsilon}$. The existence of a scalar energy function forces $\boldsymbol{\epsilon}$ to be symmetric, in other words,

$$W = \frac{1}{2}\boldsymbol{E} \cdot \boldsymbol{\epsilon} \cdot \boldsymbol{E} = \frac{1}{2}(\boldsymbol{E} \cdot \boldsymbol{\epsilon} \cdot \boldsymbol{E})^T = \frac{1}{2}\boldsymbol{E} \cdot \boldsymbol{\epsilon}^T \cdot \boldsymbol{E}, \tag{4.6}$$

which implies $\boldsymbol{\epsilon}^T = \boldsymbol{\epsilon}$. Consequently, $\boldsymbol{\epsilon}$ has only six free constants. The non-negativity of W imposes the restriction that $\boldsymbol{\epsilon}$ remains positive definite.

4.1.1 Material Symmetry

Transformation matrices are used in determining the symmetries. Consider a plane of symmetry, the $x_2 - x_3$ (Cartesian) plane (Figure 2.1). A plane of symmetry implies that the material has the same properties with respect to that plane. Therefore, we should be able to flip the axes with respect to that plane, and have no change in the constitutive law. By definition, a plane of symmetry exists at a point where the material constants have the same value for a pair of coordinate systems. The axes are referred to as "equivalent permittivity directions." Also by definition, an axis of symmetry of order K exists at a point when there are sets of equivalent permittivity directions which can be superposed by a rotation through an angle $2\pi/K$ about an axis. The way to determine permittivity symmetry is as follows. First, we rotate the vectors

$$[\hat{\boldsymbol{D}}] = [\boldsymbol{Q}][\boldsymbol{D}] \tag{4.7}$$

and

$$[\hat{\boldsymbol{E}}] = [\boldsymbol{Q}][\boldsymbol{E}], \tag{4.8}$$

and thus the transformed system is

$$[\hat{\boldsymbol{D}}] = [\hat{\boldsymbol{\epsilon}}][\hat{\boldsymbol{E}}], \tag{4.9}$$

implying

$$[\boldsymbol{D}] = [\boldsymbol{Q}]^{-1}[\hat{\epsilon}][\boldsymbol{Q}][\boldsymbol{E}] \tag{4.10}$$

where $[\boldsymbol{Q}]$ is a rotational transformation matrix. Imposing electrical symmetry means the components are invariant with respect to the transformation. Therefore,

$$[\epsilon] = [\boldsymbol{Q}^{-1}][\hat{\epsilon}][\boldsymbol{Q}] = [\boldsymbol{Q}^{-1}][\epsilon][\boldsymbol{Q}]. \tag{4.11}$$

Thus, all components that are not identical must be zero if the material has the assumed electrical symmetry. In this fashion, one can "carve" away components from a general anisotropic material tensor. The central point of such symmetries is that in a new transformed state, $\boldsymbol{\epsilon}$ should not change.

Examples of material symmetry

To clarify, consider the following steps for one plane of symmetry starting with a material having six free constants:[1]

- STEP 1: Reflect the x_1 axis

$$\boldsymbol{R}(x_1)\cdot\boldsymbol{x} \stackrel{\text{def}}{=} \begin{bmatrix} -1 & 0 & 0 \\ 0 & 1 & 0 \\ 0 & 0 & 1 \end{bmatrix} \begin{Bmatrix} x_1 \\ x_2 \\ x_3 \end{Bmatrix} = \begin{Bmatrix} \hat{x}_1 \\ \hat{x}_2 \\ \hat{x}_3 \end{Bmatrix} = \begin{Bmatrix} -x_1 \\ x_2 \\ x_3 \end{Bmatrix}. \tag{4.12}$$

- STEP 2: Transform the $\boldsymbol{E}$ and $\boldsymbol{D}$ vectors with the same transformation,

$$\boldsymbol{R}(x_1)\cdot\boldsymbol{E} \stackrel{\text{def}}{=} \begin{bmatrix} -1 & 0 & 0 \\ 0 & 1 & 0 \\ 0 & 0 & 1 \end{bmatrix} \begin{Bmatrix} E_1 \\ E_2 \\ E_3 \end{Bmatrix} = \begin{Bmatrix} \hat{E}_1 \\ \hat{E}_2 \\ \hat{E}_3 \end{Bmatrix} = \begin{Bmatrix} -E_1 \\ E_2 \\ E_3 \end{Bmatrix} \tag{4.13}$$

and

$$\boldsymbol{R}(x_1)\cdot\boldsymbol{D} \stackrel{\text{def}}{=} \begin{bmatrix} -1 & 0 & 0 \\ 0 & 1 & 0 \\ 0 & 0 & 1 \end{bmatrix} \begin{Bmatrix} D_1 \\ D_2 \\ D_3 \end{Bmatrix} = \begin{Bmatrix} \hat{D}_1 \\ \hat{D}_2 \\ \hat{D}_3 \end{Bmatrix} = \begin{Bmatrix} -D_1 \\ D_2 \\ D_3 \end{Bmatrix}. \tag{4.14}$$

- STEP 3: Transform the $\boldsymbol{\epsilon}$ with the same transformation, but for second order tensor rules:

$$\underbrace{\begin{bmatrix} -1 & 0 & 0 \\ 0 & 1 & 0 \\ 0 & 0 & 1 \end{bmatrix}}_{\boldsymbol{R}^{-1}} \begin{bmatrix} \epsilon_{11} & \epsilon_{12} & \epsilon_{13} \\ \epsilon_{21} & \epsilon_{22} & \epsilon_{23} \\ \epsilon_{31} & \epsilon_{32} & \epsilon_{33} \end{bmatrix} \underbrace{\begin{bmatrix} -1 & 0 & 0 \\ 0 & 1 & 0 \\ 0 & 0 & 1 \end{bmatrix}}_{\boldsymbol{R}} = \begin{bmatrix} \hat{\epsilon}_{11} & \hat{\epsilon}_{12} & \hat{\epsilon}_{13} \\ \hat{\epsilon}_{21} & \hat{\epsilon}_{22} & \hat{\epsilon}_{23} \\ \hat{\epsilon}_{31} & \hat{\epsilon}_{32} & \hat{\epsilon}_{33} \end{bmatrix}, \tag{4.15}$$

[1] $\boldsymbol{R}$ is a special case of $\boldsymbol{Q}$.

and thus

$$\begin{bmatrix} \hat{\epsilon}_{11} & \hat{\epsilon}_{12} & \hat{\epsilon}_{13} \\ \hat{\epsilon}_{21} & \hat{\epsilon}_{22} & \hat{\epsilon}_{23} \\ \hat{\epsilon}_{31} & \hat{\epsilon}_{32} & \hat{\epsilon}_{33} \end{bmatrix} = \begin{bmatrix} \epsilon_{11} & -\epsilon_{12} & -\epsilon_{13} \\ -\epsilon_{21} & \epsilon_{22} & \epsilon_{23} \\ -\epsilon_{31} & \epsilon_{32} & \epsilon_{33} \end{bmatrix}. \tag{4.16}$$

- STEP 4: Since the components of ϵ and $\hat{\epsilon}$ must be the same, it is a plane of symmetry, then

$$\begin{bmatrix} \epsilon_{11} & \epsilon_{12} & \epsilon_{13} \\ \epsilon_{21} & \epsilon_{22} & \epsilon_{23} \\ \epsilon_{31} & \epsilon_{32} & \epsilon_{33} \end{bmatrix} = \begin{bmatrix} \epsilon_{11} & -\epsilon_{12} & -\epsilon_{13} \\ -\epsilon_{21} & \epsilon_{22} & \epsilon_{23} \\ -\epsilon_{31} & \epsilon_{32} & \epsilon_{33} \end{bmatrix}. \tag{4.17}$$

All components that are not equal before and after the transformations (above) are zero, thus

$$\begin{Bmatrix} \hat{D}_1 \\ \hat{D}_2 \\ \hat{D}_3 \end{Bmatrix} = \begin{bmatrix} \epsilon_{11} & 0 & 0 \\ 0 & \epsilon_{22} & \epsilon_{23} \\ 0 & \epsilon_{32} & \epsilon_{33} \end{bmatrix} \begin{Bmatrix} \hat{E}_1 \\ \hat{E}_2 \\ \hat{E}_3 \end{Bmatrix}. \tag{4.18}$$

The end result is a Monoclinic material, i.e, one plane of electrical symmetry ($x_2 - x_3$ plane)(four free constants)

$$\epsilon \stackrel{\text{def}}{=} \begin{bmatrix} \epsilon_{11} & 0 & 0 \\ 0 & \epsilon_{22} & \epsilon_{23} \\ 0 & \epsilon_{32} & \epsilon_{33} \end{bmatrix}. \tag{4.19}$$

Repeating the procedure for a symmetry about the $x_1 - x_3$ plane yields

$$\epsilon \stackrel{\text{def}}{=} \begin{bmatrix} \epsilon_{11} & 0 & 0 \\ 0 & \epsilon_{22} & 0 \\ 0 & 0 & \epsilon_{33} \end{bmatrix}. \tag{4.20}$$

Repeating the procedure for a symmetry about the $x_1 - x_2$ plane changes nothing. Performing rotations about the $x_3 - axis$ yields $\epsilon_{11} = \epsilon_{22}$ and performing rotations about the $x_2 - axis$ yields $\epsilon_{11} = \epsilon_{33}$, thus yielding an isotropic material

$$\epsilon \stackrel{\text{def}}{=} \begin{bmatrix} \epsilon & 0 & 0 \\ 0 & \epsilon & 0 \\ 0 & 0 & \epsilon \end{bmatrix}. \tag{4.21}$$

$\epsilon_{11} = \epsilon_{22} = \epsilon_{33} = \epsilon$. This immediately implies that there are an infinite number of planes where the material properties are equal in all directions, and thus a single material constant. The material is of the familiar *isotropic* variety. An isotropic body has material properties that are the same in every direction at a point in the body, i.e., the properties are not a function of orientation at a point in a body. In this case, we have

$$D = \epsilon E, \tag{4.22}$$

where, under the assumption of a positive energy, we must have $\epsilon > 0$ to retain positive definiteness of $\boldsymbol{\epsilon}$. The same results hold for the magnetic fields

$$B = \mu H. \tag{4.23}$$

Remark: Obviously, the $\boldsymbol{\epsilon}$ tensor, regardless of the degree of anisotropy, must have positive eigenvalues. It is symmetric and therefore, denoting the similarity transform, the 3×3 collection of 3×1 mutually orthonormal eigenvectors of the permittivity as $\mathbf{Z}$, we have

$$[\mathbf{Z}]^T[\epsilon][\mathbf{Z}] = [\Lambda], \tag{4.24}$$

where $[\Lambda]$ is diagonalized, that is,

$$\{\boldsymbol{D}\} = [\mathbf{Z}][\boldsymbol{\Lambda}][\mathbf{Z}]^T\{\boldsymbol{E}\}, \tag{4.25}$$

implying

$$[\mathbf{Z}]^T\{\boldsymbol{D}\} = \{\hat{\boldsymbol{D}}\} = [\boldsymbol{\Lambda}][\mathbf{Z}]^T\{\boldsymbol{E}\} = [\boldsymbol{\Lambda}]\{\hat{\boldsymbol{E}}\}. \tag{4.26}$$

Clearly, the constitutive law in the new frame has the form

$$\begin{Bmatrix} \hat{D}_1 \\ \hat{D}_2 \\ \hat{D}_3 \end{Bmatrix} = \begin{bmatrix} \Lambda_1 & 0 & 0 \\ 0 & \Lambda_2 & 0 \\ 0 & 0 & \Lambda_3 \end{bmatrix} \begin{Bmatrix} \hat{E}_1 \\ \hat{E}_2 \\ \hat{E}_3 \end{Bmatrix}. \tag{4.27}$$

The energy is simply

$$W = \frac{1}{2}\left(\Lambda_1(\hat{E}_1)^2 + \Lambda_2(\hat{E}_2)^2 + \Lambda_3(\hat{E}_3)^2\right). \tag{4.28}$$

One interpretation is that of a generalized decomposition of the energy into pieces associated with the principle directions of the the permittivity tensor in matrix form. Clearly, each eigenvalue, $\Lambda_i, i = 1, 2, 3$, must be positive. Again, the same results hold for the magnetic field and corresponding magnetic energy.

5

Extraction of Macroscopic Effective Properties

5.1 Effective Properties of Heterogeneous Electromagnetic Media

In order to introduce fundamental concepts pertaining to effective properties of electromagnetic media, we initially start with static, lossless, conditions. Afterwards, we consider more general, thermally-sensitive, time-transient scenarios and the corresponding numerical methods. Let us start by simplifying the general formulations of Maxwell's equations, by simplifying Faraday's law for a static, lossless, case[1]

$$\nabla_x \times \boldsymbol{E} = -\left(\frac{\partial(\boldsymbol{\mu} \cdot \boldsymbol{H})}{\partial t} + \boldsymbol{M}^{ext} + \hat{\boldsymbol{\sigma}} \cdot \boldsymbol{H}\right) \Rightarrow \nabla_x \times \boldsymbol{E} = \boldsymbol{0} \tag{5.1}$$

and Ampere's law

$$\nabla_x \times \boldsymbol{H} = \frac{\partial(\boldsymbol{\epsilon} \cdot \boldsymbol{E})}{\partial t} + \boldsymbol{J}^{ext} + \boldsymbol{\sigma} \cdot \boldsymbol{E} \Rightarrow \nabla_x \times \boldsymbol{H} = \boldsymbol{0}, \tag{5.2}$$

where we recall

- $\boldsymbol{E}$=electric field intensity in $\frac{volts}{meter}$
- $\boldsymbol{D}$=electric flux density in $\frac{coulombs}{(meter)^2}$
- $\boldsymbol{J}$=electric current density in $\frac{amperes}{(meter)^2}$
- $\boldsymbol{H}$=magnetic field intensity in $\frac{amperes}{meter}$
- $\boldsymbol{B}$=magnetic flux density in $\frac{webers}{(meter)^2}$
- $\boldsymbol{M}$=equivalent magnetic current density in $\frac{volts}{(meter)^2}$
- ϵ=electric permittivity in $\frac{farads}{meter}$
- μ=magnetic permeability in $\frac{henrys}{meter}$

[1] Here, $\boldsymbol{M}^{ext}$ is a phenomenological "equivalent magnetic source" term and $\boldsymbol{J}^{ext}$ is a current source term.

T.I. Zohdi: Electromagnetic Properties of Multiphase Dielectrics, LNACM 64, pp. 45–62.
springerlink.com

- σ=electric conductivity in $\frac{siemens}{meter}$
- $\hat{\sigma}$=equivalent magnetic loss in $\frac{ohms}{meter}$.

Initially, we consider a simple effective property relation between $\langle \boldsymbol{D} \rangle_\Omega$ and $\langle \boldsymbol{E} \rangle_\Omega$

$$\langle \boldsymbol{D} \rangle_\Omega = \boldsymbol{\epsilon}^* \cdot \langle \boldsymbol{E} \rangle_\Omega, \tag{5.3}$$

where $\boldsymbol{\epsilon}^*$ is the effective permittivity. Similarly, we relate $\langle \boldsymbol{B} \rangle_\Omega$ and $\langle \boldsymbol{H} \rangle_\Omega$ via

$$\langle \boldsymbol{B} \rangle_\Omega = \boldsymbol{\mu}^* \cdot \langle \boldsymbol{H} \rangle_\Omega, \tag{5.4}$$

where $\boldsymbol{\mu}^*$ is the effective magnetic permeability of the medium. As stated in the introduction, it is clear that for the relation between averages to be useful, it must be computed over a sample containing a statistically representative amount of material.

5.1.1 Framing and the Construction of a Boundary Value Problem

One commonly accepted macro/micro criterion for selecting sample sizes that are used in effective property calculations is to use the well-known ergodicity condition, often referred to as Hill's condition (Hill [43], 1952)[2]

$$\langle \boldsymbol{D} \cdot \boldsymbol{E} \rangle_\Omega = \langle \boldsymbol{D} \rangle_\Omega \cdot \langle \boldsymbol{E} \rangle_\Omega. \tag{5.5}$$

Basically, this is a statement that the microelectrical-energy must equal the macroelectrical-energy. This is often referred to as an ergodicity condition in statistical mechanics. In order to understand the implications of Equation 5.5, we first split the electric field flux and electric fields into a mean part ($\langle \boldsymbol{D} \rangle_\Omega$) and a purely fluctuating part ($\tilde{\boldsymbol{D}}$), $\boldsymbol{D} = \langle \boldsymbol{D} \rangle_\Omega + \tilde{\boldsymbol{D}}$, where $\langle \tilde{\boldsymbol{D}} \rangle_\Omega = \boldsymbol{0}$ and $\boldsymbol{E} = \langle \boldsymbol{E} \rangle_\Omega + \tilde{\boldsymbol{E}}$, where $\langle \tilde{\boldsymbol{E}} \rangle_\Omega = \boldsymbol{0}$. Inserting these expressions into the system energy and expanding yields

$$\langle (\langle \boldsymbol{D} \rangle_\Omega + \tilde{\boldsymbol{D}}) \cdot (\langle \boldsymbol{E} \rangle_\Omega + \tilde{\boldsymbol{E}}) \rangle_\Omega = \langle \boldsymbol{D} \rangle_\Omega \cdot \langle \boldsymbol{E} \rangle_\Omega + \langle \tilde{\boldsymbol{D}} \cdot \tilde{\boldsymbol{E}} \rangle_\Omega \tag{5.6}$$

since $\langle \tilde{\boldsymbol{D}} \rangle_\Omega = \boldsymbol{0}$ and $\langle \tilde{\boldsymbol{E}} \rangle_\Omega = \boldsymbol{0}$. The ergodicity assumption is that $\langle \tilde{\boldsymbol{D}} \cdot \tilde{\boldsymbol{E}} \rangle_\Omega \to 0$, as the volume, $|\Omega| \to \infty$. The implication is that as the sample becomes infinitely large, $\tilde{\boldsymbol{D}} \cdot \tilde{\boldsymbol{E}}$ is purely fluctuating and hence $\langle \tilde{\boldsymbol{D}} \cdot \tilde{\boldsymbol{E}} \rangle_\Omega = 0$. In other words, the ergodicity assumption is that *the product of two purely fluctuating random fields is also purely fluctuating.* This is exactly the assertion of the Hill condition. Typically, an analyst will apply uniform loading on a large sample, with the understanding that this idealization "mimics" what a representative volume element (which is much smaller than the structural component of

[2] The following discussion holds for $\boldsymbol{B}$ and $\boldsymbol{H}$ in virtually the same manner. For more on ergodic hypotheses, see the classical work of Kröner [52].

intended use) would experience. It is clear that uniform loading is an idealization and, thus, will only be present within a vanishingly small microstructure relative to a finite-sized engineering (macro)structure.

The ergodicity assumption motivates the use of a "framing" technique (Figure 5.1), that is, a method whereby uniform far-fields are applied on the boundary of a large sample, and the interior of the sample is then probed with subsamples within the larger sample in order to avoid boundary effects that occur from imposing the uniform fields on the larger sample exterior. This is akin to exploiting a St. Venant-type of effect commonly used in solid mechanics in order to avoid boundary layers. The approach provides a way of determining what the microstructure really experiences, without "bias" from the boundary loading.

5.1.2 Computational Testing

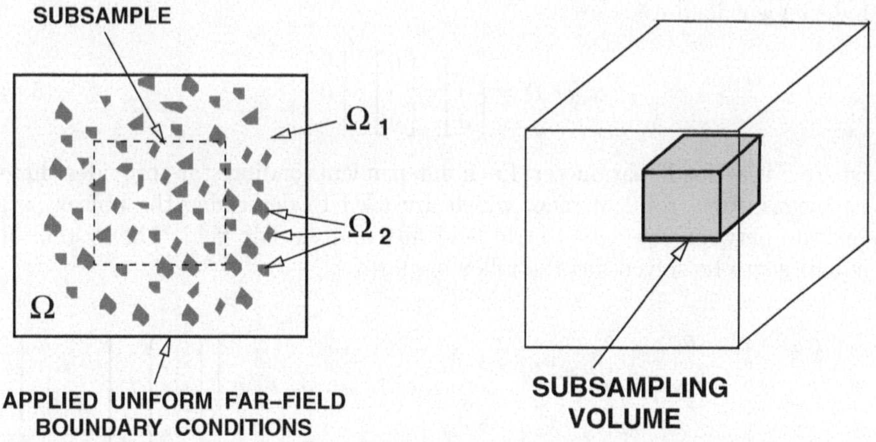

Fig. 5.1. With the framing method, a sample is probed with interior subsamples (right), within the larger sample, in order to avoid boundary effects that occur from imposing the uniform fields on the large-sample exterior (Zohdi [137]).

To determine $\boldsymbol{\epsilon}^*$, one specifies three linearly independent (uniform) loadings of the form (1) $\mathbf{E}|_{\partial\Omega} = \boldsymbol{\mathcal{E}}^{i\rightarrow iii}$ or (2) $\mathbf{D}|_{\partial\Omega} = \boldsymbol{\mathcal{D}}^{i\rightarrow iii}$, where $\boldsymbol{\mathcal{E}}^{i\rightarrow iii}$ and $\boldsymbol{\mathcal{D}}^{i\rightarrow iii}$ are vectors with spatially constant components. This loading is applied to the sample in Figure 5.1, which depicts a microheterogeneous material. Each independent loading yields three different averaged electric field components and hence provides three equations for the constitutive constants in $\boldsymbol{\epsilon}^*$. In order for such an analysis to be valid, i.e., to make the material data reliable, the sample must be small enough that it can be considered as a material point with respect to the size of the domain under analysis, but large enough to be a statistically representative sample of the microstructure. In general, in order to determine structural scale material properties of microheterogeneous material, one computes nine constitutive constants ϵ^*_{ij} in the following relation between averages (actually six due to symmetry),

$$\begin{Bmatrix} \langle D_1\rangle_\Omega \\ \langle D_2\rangle_\Omega \\ \langle D_3\rangle_\Omega \end{Bmatrix} = \begin{bmatrix} \epsilon^*_{11} & \epsilon^*_{12} & \epsilon^*_{13} \\ \epsilon^*_{21} & \epsilon^*_{22} & \epsilon^*_{23} \\ \epsilon^*_{31} & \epsilon^*_{32} & \epsilon^*_{33} \end{bmatrix} \begin{Bmatrix} \langle E_1\rangle_\Omega \\ \langle E_2\rangle_\Omega \\ \langle E_3\rangle_\Omega \end{Bmatrix}. \tag{5.7}$$

As mentioned before, each independent loading leads to three equations and, thus, in total, nine equations are generated by the three independent loadings which are used to determine the tensor relation between average electric field flux and electric field, $\boldsymbol{\epsilon}^*$. Note that $\boldsymbol{\epsilon}^*$ *is exactly what appears in engineering books as the "property" of materials.* The usual choices for the three independent load cases are

$$\boldsymbol{\mathcal{E}} \text{ or } \boldsymbol{\mathcal{D}} = \begin{bmatrix} \beta \\ 0 \\ 0 \end{bmatrix}, \begin{bmatrix} 0 \\ \beta \\ 0 \end{bmatrix}, \begin{bmatrix} 0 \\ 0 \\ \beta \end{bmatrix}, \tag{5.8}$$

where β is a "load" parameter. Each independent loading state provides three equations, for a total of nine, which are used to determine the tensor ($\boldsymbol{\epsilon}^*$) relation between average electric field flux and electric field. The system of equations to be solved has the following form:

$$\begin{bmatrix}
\langle E_1\rangle^i_\Omega & \langle E_2\rangle^i_\Omega & \langle E_3\rangle^i_\Omega & 0 & 0 & 0 & 0 & 0 & 0 \\
0 & 0 & 0 & \langle E_1\rangle^i_\Omega & \langle E_2\rangle^i_\Omega & \langle E_3\rangle^i_\Omega & 0 & 0 & 0 \\
0 & 0 & 0 & 0 & 0 & 0 & \langle E_1\rangle^i_\Omega & \langle E_2\rangle^i_\Omega & \langle E_3\rangle^i_\Omega \\
\langle E_1\rangle^{ii}_\Omega & \langle E_2\rangle^{ii}_\Omega & \langle E_3\rangle^{ii}_\Omega & 0 & 0 & 0 & 0 & 0 & 0 \\
0 & 0 & 0 & \langle E_1\rangle^{ii}_\Omega & \langle E_2\rangle^{ii}_\Omega & \langle E_3\rangle^{ii}_\Omega & 0 & 0 & 0 \\
0 & 0 & 0 & 0 & 0 & 0 & \langle E_1\rangle^{ii}_\Omega & \langle E_2\rangle^{ii}_\Omega & \langle E_3\rangle^{ii}_\Omega \\
\langle E_1\rangle^{iii}_\Omega & \langle E_2\rangle^{iii}_\Omega & \langle E_3\rangle^{iii}_\Omega & 0 & 0 & 0 & 0 & 0 & 0 \\
0 & 0 & 0 & \langle E_1\rangle^{iii}_\Omega & \langle E_2\rangle^{iii}_\Omega & \langle E_3\rangle^{iii}_\Omega & 0 & 0 & 0 \\
0 & 0 & 0 & 0 & 0 & 0 & \langle E_1\rangle^{iii}_\Omega & \langle E_2\rangle^{iii}_\Omega & \langle E_3\rangle^{iii}_\Omega
\end{bmatrix}
\begin{Bmatrix} \epsilon^*_{11} \\ \epsilon^*_{12} \\ \epsilon^*_{13} \\ \epsilon^*_{21} \\ \epsilon^*_{22} \\ \epsilon^*_{23} \\ \epsilon^*_{31} \\ \epsilon^*_{32} \\ \epsilon^*_{33} \end{Bmatrix}
= \begin{Bmatrix} \langle D_1\rangle^i_\Omega \\ \langle D_2\rangle^i_\Omega \\ \langle D_3\rangle^i_\Omega \\ \langle D_1\rangle^{ii}_\Omega \\ \langle D_2\rangle^{ii}_\Omega \\ \langle D_3\rangle^{ii}_\Omega \\ \langle D_1\rangle^{iii}_\Omega \\ \langle D_2\rangle^{iii}_\Omega \\ \langle D_3\rangle^{iii}_\Omega \end{Bmatrix}. \tag{5.9}$$

Importantly, if the effective response is assumed isotropic, then only one test loading (instead of usually three) is required, that is,

$$\epsilon^* \stackrel{\text{def}}{=} \sqrt{\frac{\langle \boldsymbol{D}\rangle_\Omega \cdot \langle \boldsymbol{D}\rangle_\Omega}{\langle \boldsymbol{E}\rangle_\Omega \cdot \langle \boldsymbol{E}\rangle_\Omega}}, \tag{5.10}$$

with a similar relation holding for

$$\mu^* \stackrel{\text{def}}{=} \sqrt{\frac{\langle \boldsymbol{B} \rangle_\Omega \cdot \langle \boldsymbol{B} \rangle_\Omega}{\langle \boldsymbol{H} \rangle_\Omega \cdot \langle \boldsymbol{H} \rangle_\Omega}}. \tag{5.11}$$

Since we will be dealing with materials comprised of randomly dispersed particulate media, we shall assume that the materials have an overall isotropic response and that Equations 5.10 and 5.11 are adequate to describe the effective material. We note that even if the aggregate response is not purely isotropic, one can interpret the above expressions as approximations of isotropic responses.

Remark 1: Applying uniform far-fields on the boundary of a large sample is a way of attempting to reproduce the length-scale disparities that are necessary for an effective property to make sense.

Remark 2: It is possible to interpret the results in another manner, which we now briefly discuss. Consider several bodies with the same external geometry composed of the same material volume fractions, but with different random microstructure. Let us focus on the same spot on each body (indexed by $i = 1, 2, ...N$) and compute

$$\frac{1}{N}\sum_{i=1}^{N} \boldsymbol{D}^{(i)} \stackrel{\text{def}}{=} \overline{\boldsymbol{\epsilon}}^{*(i)} \cdot \left(\frac{1}{N}\sum_{i=1}^{N} \boldsymbol{E}^{(i)}\right), \tag{5.12}$$

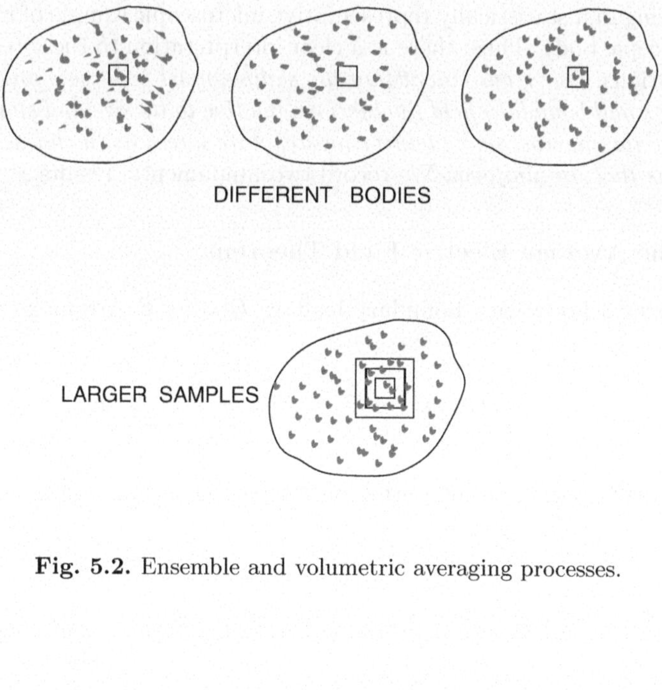

Fig. 5.2. Ensemble and volumetric averaging processes.

where N is the number of macroscopic structures being tested. The tensorial quantity, $\overline{\epsilon}^*$, is called an ensemble average (Figure 5.2). Now, we perform another experiment. We have a single body, and focus on one spot, but repeatedly compute and enlarge the sample size (indexed by $j = 1, 2, ...M$),

$$\langle \boldsymbol{D}^{(j)} \rangle_{\Omega_j} = \boldsymbol{\epsilon}^{*(j)} \cdot \langle \boldsymbol{E}^{(j)} \rangle_{\Omega_j}. \tag{5.13}$$

This is our familiar (volumetric averaged) effective property. A classical ergodicity *assumption* is that of $i, j \to \infty$ that $\overline{\epsilon}^* = \boldsymbol{\epsilon}^*$. Therefore, volumetric averaging and ensemble averaging yield the same result. In other words, *an infinitely large sample's volumetric average must equal the ensemble average of infinitely many finite samples at a point from different bodies.* Again, we refer the reader to the classical work of Kröner [52] for more details.

5.2 Averaging Theorems

For any perfectly bonded heterogeneous body, in the absence of sources, two physically important loading states are used in homogenization. They are (1) pure electric fields of the form: $\boldsymbol{E}|_{\partial\Omega} = \boldsymbol{\mathcal{E}} \Rightarrow \langle \boldsymbol{E} \rangle_\Omega = \boldsymbol{\mathcal{E}}$ and (2) pure electric field flux of the form: $\boldsymbol{D}|_{\partial\Omega} = \boldsymbol{\mathcal{D}} \Rightarrow \langle \boldsymbol{D} \rangle_\Omega = \boldsymbol{\mathcal{D}}$, where $\boldsymbol{\mathcal{E}}$ and $\boldsymbol{\mathcal{D}}$ are constant electric field and electric field flux tensors, respectively. Clearly, for these loading states to be satisfied within a macroscopic body under nonuniform external loading, the sample must be large enough to possess small boundary field fluctuations relative to its size. Therefore, applying (1)- or (2)-type boundary conditions to a large sample is a way of reproducing approximately what may be occurring in a statistically representative microscopic sample of material in a macroscopic body. Thus, there is a clear interpretation to these test boundary conditions. *Our requirement that the sample must be large enough to have relatively small boundary field fluctuations relative to its size and small enough relative to the macroscopic engineering structure forces us to choose boundary conditions that are uniform.* We record two fundamental results.

5.2.1 The Average Electric Field Theorem

We consider a body with boundary loading $\boldsymbol{E}|_{\partial\Omega} = \boldsymbol{\mathcal{E}}$. We make use of the identity

$$\nabla_x \times (\boldsymbol{E} \otimes \boldsymbol{x}) = (\nabla_x \times \boldsymbol{E}) \otimes \boldsymbol{x} + \underbrace{\boldsymbol{E} \cdot \nabla_x \boldsymbol{x}}_{\boldsymbol{E}}, \tag{5.14}$$

and substitute this in the definition of the average electric field

$$\langle \boldsymbol{E} \rangle_\Omega = \frac{1}{|\Omega|} \int_\Omega (\nabla_x \times (\boldsymbol{E} \otimes \boldsymbol{x}) - \underbrace{(\nabla_x \times \boldsymbol{E})}_{=\boldsymbol{0}} \otimes \boldsymbol{x})\, d\Omega \tag{5.15}$$

$$= \frac{1}{|\Omega|} \int_{\partial\Omega} \boldsymbol{n} \times (\boldsymbol{E} \otimes \boldsymbol{x})\, dA = \frac{1}{|\Omega|} \int_{\partial\Omega} \boldsymbol{n} \times (\boldsymbol{\mathcal{E}} \otimes \boldsymbol{x})\, dA \tag{5.16}$$

$$= \frac{1}{|\Omega|} \left(\int_\Omega (\nabla_x \times \boldsymbol{E}) \otimes \boldsymbol{x}\, d\Omega + \int_\Omega \boldsymbol{\mathcal{E}} \cdot \nabla_x \boldsymbol{x}\, d\Omega \right).$$

Thus, if $\nabla_x \times \boldsymbol{E} = \boldsymbol{0}$, then $\langle \boldsymbol{E} \rangle_\Omega = \boldsymbol{\mathcal{E}}$.

5.2.2 The Average Electric Field Flux Theorem

We consider a body with $\boldsymbol{D}|_{\partial\Omega} = \boldsymbol{\mathcal{D}}$. We make use of the identity

$$\nabla_x \cdot (\boldsymbol{D} \otimes \boldsymbol{x}) = (\nabla_x \cdot \boldsymbol{D})\boldsymbol{x} + \underbrace{\boldsymbol{D} \cdot \nabla_x \boldsymbol{x}}_{\boldsymbol{D}}, \tag{5.17}$$

and substitute this in the definition of the average electric flux

$$\langle \boldsymbol{D} \rangle_\Omega = \frac{1}{|\Omega|} \int_\Omega (\nabla_x \cdot (\boldsymbol{D} \otimes \boldsymbol{x}) - \underbrace{(\nabla_x \cdot \boldsymbol{D})}_{=0} \boldsymbol{x})\, d\Omega = \frac{1}{|\Omega|} \int_{\partial\Omega} \boldsymbol{n} \cdot (\boldsymbol{D} \otimes \boldsymbol{x})\, dA \tag{5.18}$$

$$= \frac{1}{|\Omega|} \int_{\partial\Omega} \boldsymbol{n} \cdot (\boldsymbol{\mathcal{D}} \otimes \boldsymbol{x})\, dA = \frac{1}{|\Omega|} \left(\int_\Omega (\nabla_x \cdot \boldsymbol{D}) \otimes \boldsymbol{x}\, d\Omega + \int_\Omega \boldsymbol{\mathcal{D}} \cdot \nabla_x \boldsymbol{x}\, d\Omega \right).$$

Thus, if $\nabla_x \cdot \boldsymbol{D} = 0$, then $\langle \boldsymbol{D} \rangle_\Omega = \boldsymbol{\mathcal{D}}$.

Remark 1: Generalizations of the divergence theorem have been used throughout the preceding proofs, which can be found in a variety of places, for example, Chandrasekharaiah and Debnath [10], or Malvern [64].

Remark 2: Virtually identical results hold for the magnetic fields with $\boldsymbol{H}$ taking the place of $\boldsymbol{E}$ and $\boldsymbol{B}$ taking the place of $\boldsymbol{D}$.

5.3 Effective Property Estimates

We now provide some classical analytical estimates of the effective response of heterogeneous materials.

5.3.1 The Hill–Reuss–Voigt–Wiener Bounds

The preceding results are quite important from both a theoretical and computational point of view. Until recently, the direct computation of micro-material responses was very difficult. Accordingly, classical approaches have sought to approximate or bound effective responses. Following the previous line of reasoning, many classical approaches start by splitting the electric field flux within a sample into a volume average, and a purely fluctuating part $\boldsymbol{E} = \langle \boldsymbol{E} \rangle_\Omega + \tilde{\boldsymbol{E}}$. This allows one to write

$$\begin{aligned} 0 &\le \int_\Omega \tilde{\boldsymbol{E}} \cdot \boldsymbol{\epsilon} \cdot \tilde{\boldsymbol{E}} \, d\Omega \\ &= \int_\Omega (\boldsymbol{E} \cdot \boldsymbol{\epsilon} \cdot \boldsymbol{E} - 2\langle \boldsymbol{E} \rangle_\Omega \cdot \boldsymbol{\epsilon} \cdot \boldsymbol{E} + \langle \boldsymbol{E} \rangle_\Omega \cdot \boldsymbol{\epsilon} \cdot \langle \boldsymbol{E} \rangle_\Omega) \, d\Omega \\ &= (\langle \boldsymbol{E} \rangle_\Omega \cdot \boldsymbol{\epsilon}^* \cdot \langle \boldsymbol{E} \rangle_\Omega - 2\langle \boldsymbol{E} \rangle_\Omega \cdot \langle \boldsymbol{\epsilon} \cdot \boldsymbol{E} \rangle_\Omega + \langle \boldsymbol{E} \rangle_\Omega \cdot \langle \boldsymbol{\epsilon} \rangle_\Omega \cdot \langle \boldsymbol{E} \rangle_\Omega)|\Omega| \\ &= \langle \boldsymbol{E} \rangle_\Omega \cdot (\langle \boldsymbol{\epsilon} \rangle_\Omega - \boldsymbol{\epsilon}^*) \cdot \langle \boldsymbol{E} \rangle_\Omega |\Omega|, \end{aligned} \tag{5.19}$$

where Hill's condition

$$\langle \boldsymbol{D} \rangle_\Omega \cdot \langle \boldsymbol{E} \rangle_\Omega = \langle \boldsymbol{E} \rangle_\Omega \cdot \boldsymbol{\epsilon}^* \cdot \langle \boldsymbol{E} \rangle_\Omega \tag{5.20}$$

has been used. Similarly, for the complementary case, with $\boldsymbol{D} = \langle \boldsymbol{D} \rangle_\Omega + \tilde{\boldsymbol{D}}$ and with $\langle \boldsymbol{E} \rangle_\Omega = \boldsymbol{\epsilon}^{*-1} \cdot \langle \boldsymbol{D} \rangle_\Omega$, we have

$$\begin{aligned} 0 &\le \int_\Omega \tilde{\boldsymbol{D}} \cdot \boldsymbol{\epsilon}^{-1} \cdot \tilde{\boldsymbol{D}} \, d\Omega = \int_\Omega (\boldsymbol{D} \cdot \boldsymbol{\epsilon}^{-1} \cdot \boldsymbol{D} - 2\langle \boldsymbol{D} \rangle_\Omega \cdot \boldsymbol{\epsilon}^{-1} \cdot \boldsymbol{D} + \langle \boldsymbol{D} \rangle_\Omega \cdot \boldsymbol{\epsilon}^{-1} \cdot \langle \boldsymbol{D} \rangle_\Omega) \, d\Omega \\ &= (\langle \boldsymbol{D} \rangle_\Omega \cdot \boldsymbol{\epsilon}^{*-1} \cdot \langle \boldsymbol{D} \rangle_\Omega - 2\langle \boldsymbol{E} \rangle_\Omega \cdot \langle \boldsymbol{D} \rangle_\Omega + \langle \boldsymbol{D} \rangle_\Omega \cdot \langle \boldsymbol{\epsilon}^{-1} \rangle_\Omega \cdot \langle \boldsymbol{D} \rangle_\Omega)|\Omega| \\ &= \langle \boldsymbol{D} \rangle_\Omega \cdot (\langle \boldsymbol{\epsilon}^{-1} \rangle_\Omega - \boldsymbol{\epsilon}^{*-1}) \cdot \langle \boldsymbol{D} \rangle_\Omega |\Omega|, \end{aligned} \tag{5.21}$$

where Hill's condition

$$\langle \boldsymbol{D} \rangle_\Omega \cdot \langle \boldsymbol{E} \rangle_\Omega = \langle \boldsymbol{D} \rangle_\Omega \cdot \boldsymbol{\epsilon}^{*-1} \cdot \langle \boldsymbol{D} \rangle_\Omega \tag{5.22}$$

has been used. Thus, we have

$$\langle \boldsymbol{\epsilon}^{-1} \rangle_\Omega^{-1} \le \boldsymbol{\epsilon}^* \le \langle \boldsymbol{\epsilon} \rangle_\Omega, \tag{5.23}$$

where we emphasize that this inequality means that the eigenvalues of the tensors $\boldsymbol{\epsilon}^* - \langle \boldsymbol{\epsilon}^{-1} \rangle_\Omega^{-1}$ and $\langle \boldsymbol{\epsilon} \rangle_\Omega - \boldsymbol{\epsilon}^*$ are nonnegative. Similarly, for the magnetic permeability , we have

$$\langle \boldsymbol{\mu}^{-1} \rangle_\Omega^{-1} \le \boldsymbol{\mu}^* \le \langle \boldsymbol{\mu} \rangle_\Omega. \tag{5.24}$$

The practical outcome of the analysis is that bounds on effective properties are obtained.

Remark: If a spatially uniform electric field is assumed within the RVE, $\boldsymbol{E} = \boldsymbol{E}^0$, then $\langle \boldsymbol{D} \rangle_\Omega = \langle \boldsymbol{\epsilon} \cdot \boldsymbol{E} \rangle_\Omega = \langle \boldsymbol{\epsilon} \rangle_\Omega \cdot \boldsymbol{E}^0$, which implies $\boldsymbol{\epsilon}^* = \langle \boldsymbol{\epsilon} \rangle_\Omega$. If the dual assumption of a spatially uniform electric field flux is assumed, $\boldsymbol{D} = \boldsymbol{D}^0$, leading to $\langle \boldsymbol{E} \rangle_\Omega = \langle \boldsymbol{\epsilon}^{-1} \cdot \boldsymbol{D} \rangle_\Omega = \langle \boldsymbol{\epsilon}^{-1} \rangle_\Omega \cdot \boldsymbol{D}^0$, then $\boldsymbol{\epsilon}^* = \langle \boldsymbol{\epsilon}^{-1} \rangle_\Omega^{-1}$. Equality is attained in the above bounds if the Reuss or Voigt assumptions hold, respectively.[3]

[3] Frequently, in electromagnetics, the Hill–Reuss–Voigt bounds are referred to as Wiener bounds (Wiener [109]).

5.3.2 Isotropic Cases

In the case of isotropic 2-phase materials, we have

$$\left(\frac{v_2}{\epsilon_2}+\frac{1-v_2}{\epsilon_1}\right)^{-1} = \underbrace{\langle\epsilon^{-1}(\boldsymbol{x})\rangle_\Omega^{-1}}_{if\ \boldsymbol{D}=constant} \leq \epsilon^* \leq \underbrace{\langle\epsilon(\boldsymbol{x})\rangle_\Omega}_{if\ \boldsymbol{E}=constant} = v_2\epsilon_2+(1-v_2)\epsilon_1, \tag{5.25}$$

and

$$\left(\frac{v_2}{\mu_2}+\frac{1-v_2}{\mu_1}\right)^{-1} = \underbrace{\langle\mu^{-1}(\boldsymbol{x})\rangle_\Omega^{-1}}_{if\ \boldsymbol{B}=constant} \leq \mu^* \leq \underbrace{\langle\mu(\boldsymbol{x})\rangle_\Omega}_{if\ \boldsymbol{H}=constant} = v_2\mu_2+(1-v_2)\mu_1, \tag{5.26}$$

5.3.3 The Asymptotic Hashin–Shtrikman Bounds

Improved bounds were developed in 1962 by Hashin and Shtrikman [39] based on variational principles using the concept of polarization or "filtering" of micromacro fields. Based on these formulations, they developed better asymptotic bounds on effective properties. *These bounds are sensitive to sample size and are strictly valid only asymptotically, i.e., for samples that are infinitely larger in length-scale than that of the microstructure. Hence, they are useful when the sample size is quite large relative to the microconstituent length scale.* The Hashin–Shtrikman bounds are the tightest possible bounds on isotropic effective responses, generated from isotropic microstructures where the volumetric data and phase contrasts of the constituents are the only data known. For the overall permittivity,

$$\langle\epsilon^{-1}(\boldsymbol{x})\rangle_\Omega^{-1} \leq \underbrace{\epsilon_1+\frac{v_2^\epsilon}{\frac{1}{\epsilon_2-\epsilon_1}+\frac{1-v_2^\epsilon}{3\epsilon_1}}}_{\epsilon^{*,-}} \leq \epsilon^* \leq \underbrace{\epsilon_2+\frac{1-v_2^\epsilon}{\frac{1}{\epsilon_1-\epsilon_2}+\frac{v_2^\epsilon}{3\epsilon_2}}}_{\epsilon^{*,+}} \leq \langle\epsilon(\boldsymbol{x})\rangle_\Omega \tag{5.27}$$

and the overall permeability,

$$\langle\mu^{-1}(\boldsymbol{x})\rangle_\Omega^{-1} \leq \underbrace{\mu_1+\frac{v_2}{\frac{1}{\mu_2-\mu_1}+\frac{1-v_2}{3\mu_1}}}_{\mu^{*,-}} \leq \mu^* \leq \underbrace{\mu_2+\frac{1-v_2}{\frac{1}{\mu_1-\mu_2}+\frac{v_2}{3\mu_2}}}_{\mu^{*,+}} \leq \langle\mu(\boldsymbol{x})\rangle_\Omega, \tag{5.28}$$

where $\epsilon_2 \geq \epsilon_1$, $\mu_2 \geq \mu_1$, v_2^ϵ is the volume fraction of phase with the higher ϵ value ("phase 2" in the former expression) for the permittivity-mismatch and v_2 is the volume fraction of the phase with the higher μ value ("phase 2" in the latter expression) for the permeability-mismatch. [4] Such bounds are the tightest possible on isotropic effective responses, with isotropic two phase microstructures, where only the volume fractions and phase contrasts of the constituents are known. Note that no further geometric information, for example, the number and nature of particles, etc., contributes to these bounds.

Remark 1: There exist a multitude of other approaches which seek to estimate or bound the aggregate responses microheterogeneous materials. A complete survey is outside the scope of the present work. We refer the reader to the works of Hashin [42], Mura [72], Nemat–Nasser and Hori [74] and recently Torquato [104] for such reviews.

Remark 2: The typical behavior of the bounds and their range, plotted against volume fraction, is shown in Figures 5.3 and 5.4. Typically, dielectric materials are measured relative to the vacuum-level permittivity, $\epsilon_o = 8.854 \times 10^{-12}$ farads/meter. The usual representation is $\epsilon_{ir} \stackrel{\text{def}}{=} \frac{\epsilon_i}{\epsilon_o}$, where ϵ_{ir} is known as the relative permittivity. [5] In this example, we chose a mixture of Polyethylene ($\epsilon_{1r} = \frac{\epsilon_1}{\epsilon_o} = 2.25$) and Silicon ($\epsilon_{2r} = \frac{\epsilon_2}{\epsilon_o} = 11.68$), both measured at Standard Temperature and Pressure, for 0.9 MHz (Hector [44]).

Remark 3: From now on, we assume that $v_2^{\epsilon} = v_2^{\mu} = v_2$.

5.4 Phase Concentration and Load-Sharing

It is sometimes useful to extract the proportion (load-share) of each field carried by the phases in the material. There are relatively fast (classical) ways to do this, with knowledge of only the effective properties, the properties of each phase and the respective volume fractions. Towards this end, consider the following identities:

$$\langle \boldsymbol{E} \rangle_{\Omega} = \frac{1}{|\Omega|}(\int_{\Omega_1} \boldsymbol{E}\, d\Omega + \int_{\Omega_2} \boldsymbol{E}\, d\Omega) = v_1 \langle \boldsymbol{E} \rangle_{\Omega_1} + v_2 \langle \boldsymbol{E} \rangle_{\Omega_2} \tag{5.29}$$

and

$$\langle \boldsymbol{D} \rangle_{\Omega} = \frac{1}{|\Omega|}(\int_{\Omega_1} \boldsymbol{D}\, d\Omega + \int_{\Omega_2} \boldsymbol{D}\, d\Omega) = v_1 \langle \boldsymbol{D} \rangle_{\Omega_1} + v_2 \langle \boldsymbol{D} \rangle_{\Omega_2}. \tag{5.30}$$

By direct manipulation, we obtain

[4] For either case, the volume fraction of the other phase is v_1, where $v_1 + v_2 = 1$.

[5] Similar representations can be made for the magnetic properties using the vacuum-level permeability , $\mu_o = 8.854 \times 10^{-12}$ farads/meter and $\mu_{ir} \stackrel{\text{def}}{=} \frac{\mu_i}{\mu_o}$, where μ_{ir} is known as the relative permeability.

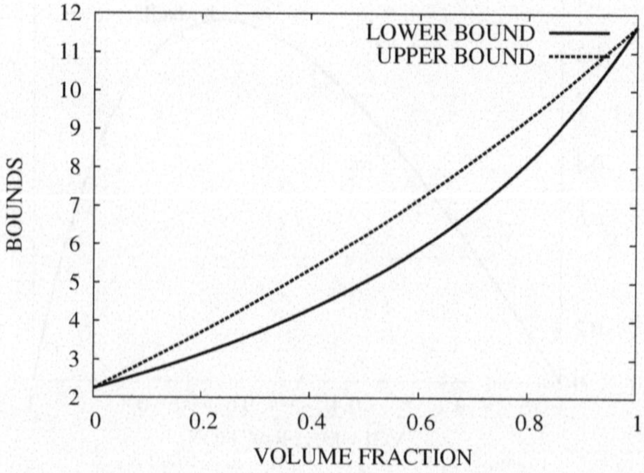

Fig. 5.3. An example of the Hashin–Shtrikman bounds for a mixture of Polyethylene ($\epsilon_{1r} = \frac{\epsilon_1}{\epsilon_o} = 2.25$) and Silicon ($\epsilon_{2r} = \frac{\epsilon_2}{\epsilon_o} = 11.68$), both measured at STP for 0.9 MHz (Hector [44]).

$$\begin{aligned}\langle \boldsymbol{D}\rangle_\Omega &= v_1\langle \boldsymbol{D}\rangle_{\Omega_1} + v_2\langle \boldsymbol{D}\rangle_{\Omega_2}\\ &= v_1\boldsymbol{\epsilon}_1\cdot\langle \boldsymbol{E}\rangle_{\Omega_1} + v_2\boldsymbol{\epsilon}_2\cdot\langle \boldsymbol{E}\rangle_{\Omega_2}\\ &= \boldsymbol{\epsilon}_1\cdot(\langle \boldsymbol{E}\rangle_\Omega - v_2\langle \boldsymbol{E}\rangle_{\Omega_2}) + v_2\boldsymbol{\epsilon}_2\cdot\langle \boldsymbol{E}\rangle_{\Omega_2}\\ &= \underbrace{(\boldsymbol{\epsilon}_1 + v_2(\boldsymbol{\epsilon}_2-\boldsymbol{\epsilon}_1)\cdot\boldsymbol{C}_{E,2})}_{\boldsymbol{\epsilon}^*}\cdot\langle \boldsymbol{E}\rangle_\Omega \end{aligned} \tag{5.31}$$

where

$$\underbrace{\left(\frac{1}{v_2}(\boldsymbol{\epsilon}_2-\boldsymbol{\epsilon}_1)^{-1}\cdot(\boldsymbol{\epsilon}^*-\boldsymbol{\epsilon}_1)\right)}_{\stackrel{\text{def}}{=}\boldsymbol{C}_{E,2}}\cdot\langle \boldsymbol{E}\rangle_\Omega = \langle \boldsymbol{E}\rangle_{\Omega_2}. \tag{5.32}$$

In the special case of isotropy,

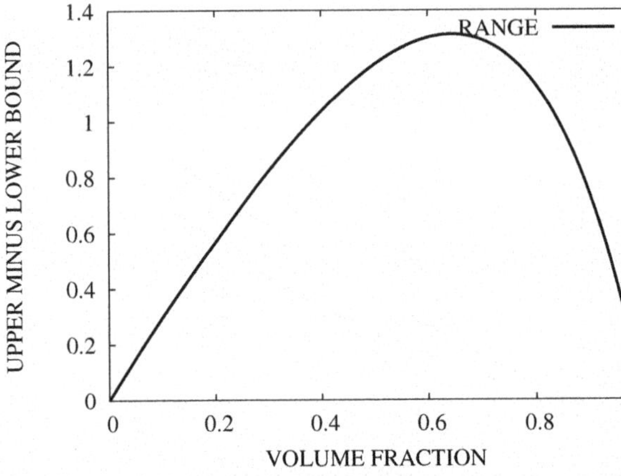

Fig. 5.4. An example of the *difference/range* in the upper and lower Hashin–Shtrikman bounds, $R_{\epsilon^*} = \frac{\epsilon^{*,+} - \epsilon^{*,-}}{\epsilon_o}$, for a mixture of Polyethylene ($\epsilon_{1r} = \frac{\epsilon_1}{\epsilon_o} = 2.25$) and Silicon ($\epsilon_{2r} = \frac{\epsilon_2}{\epsilon_o} = 11.68$), both measured at STP for 0.9 MHz (Hector [44]).

$$C_{E,2} \stackrel{\text{def}}{=} \frac{1}{v_2} \frac{\epsilon^* - \epsilon_1}{\epsilon_2 - \epsilon_1}. \tag{5.33}$$

Thereafter, we may write, for the variation in the electric field flux,

$$\boldsymbol{C}_{E,2} \cdot \boldsymbol{\epsilon}^{*-1} \cdot \langle \boldsymbol{D} \rangle_{\Omega} = \boldsymbol{\epsilon}_2^{-1} \cdot \langle \boldsymbol{D} \rangle_{\Omega_2}, \tag{5.34}$$

which reduces to

$$\boldsymbol{\epsilon}_2 \cdot \boldsymbol{C}_{E,2} \cdot \boldsymbol{\epsilon}^{*-1} \cdot \langle \boldsymbol{D} \rangle_{\Omega} \stackrel{\text{def}}{=} \boldsymbol{C}_{D,2} \cdot \langle \boldsymbol{D} \rangle_{\Omega} = \langle \boldsymbol{D} \rangle_{\Omega_2}. \tag{5.35}$$

$\boldsymbol{C}_{D,2}$ is known as the electric field flux concentration tensor. Therefore, once either $\boldsymbol{C}_{D,2}$ or $\boldsymbol{\epsilon}^*$ are known, the other can be determined. In the case of isotropy, we may write

$$C_{D,2} \stackrel{\text{def}}{=} \frac{1}{v_2} \frac{\epsilon_2}{\epsilon^*} \frac{\epsilon^* - \epsilon_1}{\epsilon_2 - \epsilon_1}. \tag{5.36}$$

For the matrix,

$$\langle \boldsymbol{E} \rangle_{\Omega_1} = \frac{\langle \boldsymbol{E} \rangle_{\Omega} - v_2 \langle \boldsymbol{E} \rangle_{\Omega_2}}{v_1} = \frac{(\mathbf{1} - v_2 \boldsymbol{C}_{E,2}) \cdot \langle \boldsymbol{E} \rangle_{\Omega}}{v_1} \stackrel{\text{def}}{=} \boldsymbol{C}_{E,1} \cdot \langle \boldsymbol{E} \rangle_{\Omega}, \quad (5.37)$$

where

$$\boldsymbol{C}_{E,1} \stackrel{\text{def}}{=} \frac{1}{v_1} (\mathbf{1} - v_2 \boldsymbol{C}_{E,2}) = \frac{\mathbf{1} - v_2 \boldsymbol{C}_{E,2}}{1 - v_2}, \quad (5.38)$$

where in the case of isotropy,

$$C_{E,1} = \frac{1 - v_2 C_{E,2}}{1 - v_2}. \quad (5.39)$$

Similarly, for the electric field flux,

$$\boldsymbol{C}_{D,1} \cdot \langle \boldsymbol{D} \rangle_{\Omega} = \langle \boldsymbol{D} \rangle_{\Omega_1}, \quad (5.40)$$

where

$$\boldsymbol{C}_{D,1} = \frac{\mathbf{1} - v_2 \boldsymbol{C}_{D,2}}{1 - v_2}, \quad (5.41)$$

where in the case of isotropy

$$C_{D,1} = \frac{1 - v_2 C_{D,2}}{1 - v_2}. \quad (5.42)$$

The exact procedure holds for the magnetic fields, with $\boldsymbol{H}$ replacing $\boldsymbol{E}$, $\boldsymbol{B}$ replacing $\boldsymbol{D}$, $\boldsymbol{\mu}$ replacing $\boldsymbol{\epsilon}$, v_1 replacing v_1 and v_2 replacing v_2. In summary, we have the following concentration tensors:

- $\boldsymbol{C}_{E,2} \cdot \langle \boldsymbol{E} \rangle_{\Omega} = \langle \boldsymbol{E} \rangle_{\Omega_2}$ where $\boldsymbol{C}_{E,2} = \frac{1}{v_2} (\boldsymbol{\epsilon}_2 - \boldsymbol{\epsilon}_1)^{-1} \cdot (\boldsymbol{\epsilon}^* - \boldsymbol{\epsilon}_1)$
- $\boldsymbol{C}_{E,1} \cdot \langle \boldsymbol{E} \rangle_{\Omega} = \langle \boldsymbol{E} \rangle_{\Omega_1}$ where $\boldsymbol{C}_{E,1} = \frac{1}{v_1} (\mathbf{1} - v_2 \boldsymbol{C}_{E,2}) = \frac{\mathbf{1} - v_2 \boldsymbol{C}_{E,2}}{1 - v_2}$
- $\boldsymbol{C}_{D,2} \cdot \langle \boldsymbol{D} \rangle_{\Omega} = \langle \boldsymbol{D} \rangle_{\Omega_2}$ where $\boldsymbol{C}_{D,2} = \boldsymbol{\epsilon}_2 \cdot \boldsymbol{C}_{E,2} \cdot \boldsymbol{\epsilon}^{*-1}$
- $\boldsymbol{C}_{D,1} \cdot \langle \boldsymbol{D} \rangle_{\Omega} = \langle \boldsymbol{D} \rangle_{\Omega_1}$ where $\boldsymbol{C}_{D,1} = \frac{\mathbf{1} - v_2 \boldsymbol{C}_{D,2}}{1 - v_2}$
- $\boldsymbol{C}_{H,2} \cdot \langle \boldsymbol{H} \rangle_{\Omega} = \langle \boldsymbol{H} \rangle_{\Omega_2}$ where $\boldsymbol{C}_{H,2} = \frac{1}{v_2} (\boldsymbol{\mu}_2 - \boldsymbol{\mu}_1)^{-1} \cdot (\boldsymbol{\mu}^* - \boldsymbol{\mu}_1)$
- $\boldsymbol{C}_{H,1} \cdot \langle \boldsymbol{H} \rangle_{\Omega} = \langle \boldsymbol{H} \rangle_{\Omega_1}$ where $\boldsymbol{C}_{H,1} = \frac{1}{v_1} (\mathbf{1} - v_2 \boldsymbol{C}_{H,2}) = \frac{\mathbf{1} - v_2 \boldsymbol{C}_{H,2}}{1 - v_2}$
- $\boldsymbol{C}_{B,2} \cdot \langle \boldsymbol{B} \rangle_{\Omega} = \langle \boldsymbol{B} \rangle_{\Omega_2}$ where $\boldsymbol{C}_{B,2} = \boldsymbol{\mu}_2 \cdot \boldsymbol{C}_{B,2} \cdot \boldsymbol{\mu}^{*-1}$
- $\boldsymbol{C}_{B,1} \cdot \langle \boldsymbol{B} \rangle_{\Omega} = \langle \boldsymbol{B} \rangle_{\Omega_1}$ where $\boldsymbol{C}_{B,1} = \frac{\mathbf{1} - v_2 \boldsymbol{C}_{B,2}}{1 - v_2}$.

Remark: *The concentration tensors indicate the amplification of the field within the particle relative to the average of the field. There has been no approximation yet.* The "burden" in the computations has shifted to the determination of the $\boldsymbol{C}$'s. Classical methods approximate them. For example, the simplest approximation is $\boldsymbol{C}_{E,2} = \mathbf{1}$, which is an approximation assuming a constant $\boldsymbol{E}$-field throughout the microstructure, while the dual approximation is $\boldsymbol{C}_{D,2} = \mathbf{1}$, corresponding to a constant $\boldsymbol{D}$-field throughout the microstructure.

5.4.1 "Load Sharing" Interpretation

Directly from Equations 5.45 and 5.46, one may write

$$
\begin{aligned}
\underbrace{v_1 \boldsymbol{C}_{E,1}}_{\text{phase}-1 \text{ contribution}} + \underbrace{v_2 \boldsymbol{C}_{E,2}}_{\text{phase}-2 \text{ contribution}} &= \mathbf{1} \\
\underbrace{v_1 \boldsymbol{C}_{D,1}}_{\text{phase}-1 \text{ contribution}} + \underbrace{v_2 \boldsymbol{C}_{D,2}}_{\text{phase}-2 \text{ contribution}} &= \mathbf{1} \\
\underbrace{v_1 \boldsymbol{C}_{H,1}}_{\text{phase}-1 \text{ contribution}} + \underbrace{v_2 \boldsymbol{C}_{H,2}}_{\text{phase}-2 \text{ contribution}} &= \mathbf{1} \\
\underbrace{v_1 \boldsymbol{C}_{B,1}}_{\text{phase}-1 \text{ contribution}} + \underbrace{v_2 \boldsymbol{C}_{B,2}}_{\text{phase}-2 \text{ contribution}} &= \mathbf{1}.
\end{aligned}
\tag{5.43}
$$

Frequently, the first term in the above expressions is referred to as "phase-1's" load share, while the second term is "phase-2's" load share.

5.5 Current-Fields

Another field-quantity that is import to keep track of, in an overall and phase-wise sense, is the current field. Many electronic devices consist of an easily moldable matrix material which is doped with particulates with different dielectric constants in order to tailor the overall dielectric response of the material for a specific application (Figure 5.5). In the context of electrical materials, the microscale properties are characterized by a spatially variable electrical conductivity $\boldsymbol{\sigma}(\boldsymbol{x})$. Typically, in order to characterize the structural-scale effective response of such materials, a relation between volume averages

$$\langle \boldsymbol{J} \rangle_\Omega = \boldsymbol{\sigma}^* \cdot \langle \boldsymbol{E} \rangle_\Omega \tag{5.44}$$

is formed where $\langle \cdot \rangle_\Omega \overset{\text{def}}{=} \frac{1}{|\Omega|} \int_\Omega \cdot \, d\Omega$ is the volume averaging operator, $\boldsymbol{J}$ is the current and $\boldsymbol{E}$ is the electric field within a statistically representative volume element (RVE) of volume $|\Omega|$. The quantity, $\boldsymbol{\sigma}^*$, is known as the effective conductivity, and is the property used in usual (homogenized) macroscale analyses. Current fields are important to ascertain because they can lead to localized overheating, i.e., "hot spots."

Accordingly, we consider a two-phase medium. One can determine the electric field "load level" carried by each phase by considering the following identities:

$$\langle \boldsymbol{E} \rangle_\Omega = \frac{1}{|\Omega|}\left(\int_{\Omega_1} \boldsymbol{E} \, d\Omega + \int_{\Omega_2} \boldsymbol{E} \, d\Omega\right) = v_1 \langle \boldsymbol{E} \rangle_{\Omega_1} + v_2 \langle \boldsymbol{E} \rangle_{\Omega_2} \tag{5.45}$$

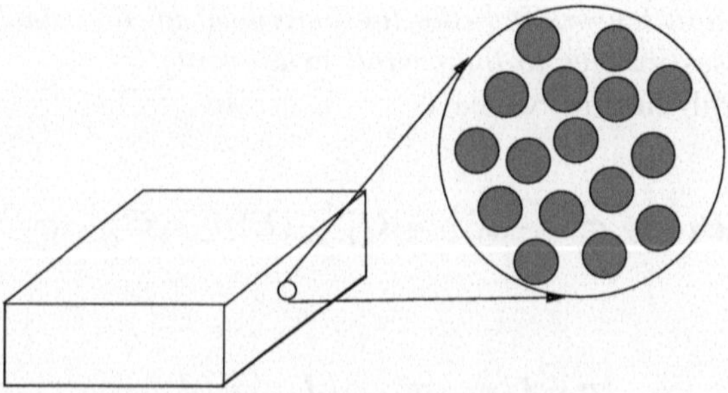

Fig. 5.5. A material with particulate additives (Zohdi [138]).

and

$$\langle \boldsymbol{J}\rangle_{\Omega} = \frac{1}{|\Omega|}\left(\int_{\Omega_1} \boldsymbol{J}\, d\Omega + \int_{\Omega_2} \boldsymbol{J}\, d\Omega\right) = v_1\langle \boldsymbol{J}\rangle_{\Omega_1} + v_2\langle \boldsymbol{J}\rangle_{\Omega_2}, \tag{5.46}$$

where v_1 and v_2 are the volume fractions of phases 1 and 2 respectively, so that $v_1 + v_2 = 1$. Performing straightforward algebraic manipulations yields

$$\begin{aligned}
\langle \boldsymbol{J}\rangle_{\Omega} &= v_1\langle \boldsymbol{J}\rangle_{\Omega_1} + v_2\langle \boldsymbol{J}\rangle_{\Omega_2} \\
&= v_1\boldsymbol{\sigma}_1 \cdot \langle \boldsymbol{E}\rangle_{\Omega_1} + v_2\boldsymbol{\sigma}_2 \cdot \langle \boldsymbol{E}\rangle_{\Omega_2} \\
&= \boldsymbol{\sigma}_1 \cdot (\langle \boldsymbol{E}\rangle_{\Omega} - v_2\langle \boldsymbol{E}\rangle_{\Omega_2}) + v_2\boldsymbol{\sigma}_2 \cdot \langle \boldsymbol{E}\rangle_{\Omega_2} \\
&= \underbrace{(\boldsymbol{\sigma}_1 + v_2(\boldsymbol{\sigma}_2 - \boldsymbol{\sigma}_1) \cdot \boldsymbol{C}_{E,2})}_{\boldsymbol{\sigma}^*} \cdot \langle \boldsymbol{E}\rangle_{\Omega},
\end{aligned} \tag{5.47}$$

where

$$\underbrace{\left(\frac{1}{v_2}(\boldsymbol{\sigma}_2 - \boldsymbol{\sigma}_1)^{-1} \cdot (\boldsymbol{\sigma}^* - \boldsymbol{\sigma}_1)\right)}_{\stackrel{\text{def}}{=}\boldsymbol{C}_{E,2}} \cdot \langle \boldsymbol{E}\rangle_{\Omega} = \langle \boldsymbol{E}\rangle_{\Omega_2}. \tag{5.48}$$

$\boldsymbol{C}_{E,2}$ is known as the electric field concentration tensor. *Once either $\boldsymbol{C}_{E,2}$ or $\boldsymbol{\sigma}^*$ is known, the other can be determined.* From Equation 5.45, we have

$$\langle \boldsymbol{E}\rangle_{\Omega_1} = \frac{\langle \boldsymbol{E}\rangle_{\Omega} - v_2\langle \boldsymbol{E}\rangle_{\Omega_2}}{v_1} = \frac{(\boldsymbol{1} - v_2\boldsymbol{C}_{E,2}) \cdot \langle \boldsymbol{E}\rangle_{\Omega}}{v_1} \stackrel{\text{def}}{=} \boldsymbol{C}_{E,1} \cdot \langle \boldsymbol{E}\rangle_{\Omega}, \tag{5.49}$$

where

$$\boldsymbol{C}_{E,1} = \frac{1}{v_1}(\mathbf{1} - v_2\boldsymbol{C}_{E,2}) = \frac{\mathbf{1} - v_2\boldsymbol{C}_{E,2}}{1 - v_2}. \tag{5.50}$$

The concentration tensors indicate the amplification or reduction of the field within the phases relative to the overall field average.

For the overall current, we have

$$\langle \boldsymbol{J}\rangle_\Omega = \boldsymbol{\sigma}^* \cdot \langle \boldsymbol{E}\rangle_\Omega \Rightarrow \boldsymbol{\sigma}^{*-1} \cdot \langle \boldsymbol{J}\rangle_\Omega = \boldsymbol{C}_{E,2}^{-1} \cdot \langle \boldsymbol{E}\rangle_{\Omega_2} = \boldsymbol{C}_{E,2}^{-1} \cdot \boldsymbol{\sigma}_2^{-1} \cdot \langle \boldsymbol{J}\rangle_{\Omega_2}. \tag{5.51}$$

Thus,

$$\underbrace{\boldsymbol{\sigma}_2 \cdot \boldsymbol{C}_{E,2} \cdot \boldsymbol{\sigma}^{*-1}}_{\boldsymbol{C}_{J,2}} \cdot \langle \boldsymbol{J}\rangle_\Omega = \langle \boldsymbol{J}\rangle_{\Omega_2}, \tag{5.52}$$

and

$$\boldsymbol{C}_{J,1} \cdot \langle \boldsymbol{J}\rangle_\Omega = \langle \boldsymbol{J}\rangle_{\Omega_1} \tag{5.53}$$

where

$$\boldsymbol{C}_{J,1} = \frac{\mathbf{1} - v_2\boldsymbol{C}_{J,2}}{1 - v_2} = \boldsymbol{\sigma}_1 \cdot \boldsymbol{C}_{E,1} \cdot \boldsymbol{\sigma}^{*-1}. \tag{5.54}$$

In summary, we have the following concentration tensors:

- $\boldsymbol{C}_{E,2} \cdot \langle \boldsymbol{E}\rangle_\Omega = \langle \boldsymbol{E}\rangle_{\Omega_2}$ where $\boldsymbol{C}_{E,2} = \frac{1}{v_2}(\boldsymbol{\sigma}_2 - \boldsymbol{\sigma}_1)^{-1} \cdot (\boldsymbol{\sigma}^* - \boldsymbol{\sigma}_1)$,
- $\boldsymbol{C}_{E,1} \cdot \langle \boldsymbol{E}\rangle_\Omega = \langle \boldsymbol{E}\rangle_{\Omega_1}$ where $\boldsymbol{C}_{E,1} = \frac{1}{v_1}(\mathbf{1} - v_2\boldsymbol{C}_{E,2}) = \frac{\mathbf{1}-v_2\boldsymbol{C}_{E,2}}{1-v_2}$,
- $\boldsymbol{C}_{J,2} \cdot \langle \boldsymbol{J}\rangle_\Omega = \langle \boldsymbol{J}\rangle_{\Omega_2}$ where $\boldsymbol{C}_{J,2} = \boldsymbol{\sigma}_2 \cdot \boldsymbol{C}_{E,2} \cdot \boldsymbol{\sigma}^{*-1}$,
- $\boldsymbol{C}_{J,1} \cdot \langle \boldsymbol{J}\rangle_\Omega = \langle \boldsymbol{J}\rangle_{\Omega_1}$ where $\boldsymbol{C}_{J,1} = \frac{\mathbf{1}-v_2\boldsymbol{C}_{J,2}}{1-v_2} = \boldsymbol{\sigma}_1 \cdot \boldsymbol{C}_{E,1} \cdot \boldsymbol{\sigma}^{*-1}$.

Remark 1: Similar to the previous results for $\boldsymbol{E}$ and $\boldsymbol{D}$, directly from Equations 5.45 and 5.46, one may write

$$\begin{aligned} \underbrace{v_1\boldsymbol{C}_{E,1}}_{\text{phase}-1 \text{ contribution}} + \underbrace{v_2\boldsymbol{C}_{E,2}}_{\text{phase}-2 \text{ contribution}} &= \mathbf{1}, \\ \underbrace{v_1\boldsymbol{C}_{J,1}}_{\text{phase}-1 \text{ contribution}} + \underbrace{v_2\boldsymbol{C}_{J,2}}_{\text{phase}-2 \text{ contribution}} &= \mathbf{1}. \end{aligned} \tag{5.55}$$

Remark 2: If the overall property is isotropic, as well as each of the constituents (for example, a microstructure comprised of an isotropic binder embedded with randomly distributed isotropic particles), then we have $\boldsymbol{C}_{E,i} = C_{E,i}\mathbf{1}$, where, explicitly for a two-phase material,

$$C_{E,1} = \frac{1}{1 - v_2}\frac{\sigma_2 - \sigma^*}{\sigma_2 - \sigma_1}, \tag{5.56}$$

and

$$C_{E,2} = \frac{1}{v_2}\frac{\sigma^* - \sigma_1}{\sigma_2 - \sigma_1}, \tag{5.57}$$

and $\boldsymbol{C}_{J,i} = C_{J,i}\mathbf{1}$, leading to

$$C_{J,1} = \frac{\sigma_1}{\sigma^*(1 - v_2)}\left(\frac{\sigma_2 - \sigma^*}{\sigma_2 - \sigma_1}\right) \tag{5.58}$$

$$C_{J,2} = \frac{\sigma_2}{\sigma^* v_2}\left(\frac{\sigma^* - \sigma_1}{\sigma_2 - \sigma_1}\right). \tag{5.59}$$

Remark 3: Recall, that the overall volume average of the electric field $\langle \boldsymbol{E}\rangle_\Omega$ in a RVE sample is equivalent to the loading on the boundary for uniform loading conditions.

5.5.1 Remark: Averaging Theorems

Two physically important (boundary) loading states satisfy the micromacro energy/power condition in Equation 6.3. They are (1) pure electric fields of the form: $\boldsymbol{E}|_{\partial\Omega} = \boldsymbol{\mathcal{E}} \Rightarrow \langle \boldsymbol{E}\rangle_\Omega = \boldsymbol{\mathcal{E}}$ and (2) pure current fields of the form: $\boldsymbol{J}|_{\partial\Omega} = \boldsymbol{\mathcal{J}} \Rightarrow \langle \boldsymbol{J}\rangle_\Omega = \boldsymbol{\mathcal{J}}$, where $\boldsymbol{\mathcal{E}}$ and $\boldsymbol{\mathcal{J}}$ are constant electric field and current field vectors, respectively. Clearly, for these loading states to be satisfied within a macroscopic body under nonuniform external loading, the sample must be large enough to possess small boundary field fluctuations relative to its size. Therefore, applying (1)- or (2)-type boundary conditions to a large sample is a way of approximately reproducing what may be occurring in a statistically representative microscopic sample of material within a macroscopic body.

5.6 Remarks

In closing, we record two important observations.

5.6.1 The Average Current Theorem

We consider a body with $\boldsymbol{J}|_{\partial\Omega} = \boldsymbol{\mathcal{J}}$. We make use of the identity

$$\nabla \cdot (\boldsymbol{J} \otimes \boldsymbol{x}) = (\nabla \cdot \boldsymbol{J})\boldsymbol{x} + \underbrace{\boldsymbol{J} \cdot \nabla \boldsymbol{x}}_{\boldsymbol{J}}, \tag{5.60}$$

and substitute this in the definition of the average current

$$\langle \boldsymbol{J}\rangle_\Omega = \frac{1}{|\Omega|}\int_\Omega (\nabla \cdot (\boldsymbol{J} \otimes \boldsymbol{x}) - \underbrace{(\nabla \cdot \boldsymbol{J})\boldsymbol{x}}_{=0})\, d\Omega = \frac{1}{|\Omega|}\int_{\partial\Omega} \boldsymbol{n} \cdot (\boldsymbol{J} \otimes \boldsymbol{x})\, dA \tag{5.61}$$

$$= \frac{1}{|\Omega|}\int_{\partial\Omega} \boldsymbol{n} \cdot (\boldsymbol{\mathcal{J}} \otimes \boldsymbol{x})\, dA = \frac{1}{|\Omega|}\left(\int_\Omega (\nabla \cdot \boldsymbol{\mathcal{J}}) \otimes \boldsymbol{x}\, d\Omega + \int_\Omega \boldsymbol{\mathcal{J}} \cdot \nabla \boldsymbol{x}\, d\Omega\right).$$

Thus, if $\nabla \cdot \boldsymbol{J} = 0$, then $\langle \boldsymbol{J}\rangle_\Omega = \boldsymbol{\mathcal{J}}$.

5.6.2 Special Cases of Maxwell's Equations

By taking the divergence of Ampere's law:

$$\nabla \cdot \left(\nabla \times \boldsymbol{H} - \frac{\partial \boldsymbol{D}}{\partial t} - \boldsymbol{J}^{ext} - \boldsymbol{J} \right) = 0, \tag{5.62}$$

one obtains

$$\nabla \cdot (\nabla \times \boldsymbol{H}) = 0, \tag{5.63}$$

$$\nabla \cdot \left(\frac{\partial \boldsymbol{D}}{\partial t} + \boldsymbol{J}^{ext} + \boldsymbol{J} \right) = 0, \tag{5.64}$$

and thus

$$\frac{\partial}{\partial t} \underbrace{\nabla \cdot \boldsymbol{D}}_{\mathcal{P}} + \nabla \cdot (\boldsymbol{J}^{ext} + \boldsymbol{J}) = 0, \tag{5.65}$$

and thus

$$\frac{\partial \mathcal{P}}{\partial t} + \nabla \cdot (\boldsymbol{J}^{ext} + \boldsymbol{J}) = 0. \tag{5.66}$$

Thus, if $\mathcal{P} = 0$ and $\nabla \cdot \boldsymbol{J}^{ext} = 0$,

$$\nabla \cdot \boldsymbol{J} = 0. \tag{5.67}$$

If one employs the constitutive relation $\boldsymbol{J} = \boldsymbol{\sigma} \cdot \boldsymbol{E}$, then this allows for Hashin–Shtrikman type estimates for effective conductivity, as does the PDE $\nabla \cdot \boldsymbol{D} = 0$ (which is valid only then $\mathcal{P} = 0$) for the effective permittivity, $\langle \boldsymbol{D} \rangle_{\Omega} = \boldsymbol{\epsilon}^* \cdot \langle \boldsymbol{E} \rangle_{\Omega}$, when $\boldsymbol{D} = \boldsymbol{\epsilon} \cdot \boldsymbol{E}$. For example, one case when these two physical situations are compatible is when

$$\boldsymbol{E} = \boldsymbol{\sigma}^{-1} \cdot \boldsymbol{J} = \boldsymbol{\epsilon}^{-1} \cdot \boldsymbol{D} \Rightarrow \boldsymbol{J} = (\boldsymbol{\sigma} \cdot \boldsymbol{\epsilon}^{-1}) \cdot \boldsymbol{D}. \tag{5.68}$$

6

Coupled Effects: Joule-Heating

6.1 Introduction

Heterogeneous microstructures lead to a distortion of the electrical and current field within the material mixture. This leads to the fields becoming amplified within the material, which can lead to a variety of detrimental effects. An important quantity of interest is the amount of heat generated from an electrical field. The interconversions of various forms of energy (electromagnetic, thermal, etc.) in a system are governed by the first law of thermodynamics (which will be derived in detail shortly),

$$\rho \dot{w} - \boldsymbol{T} : \nabla \dot{\boldsymbol{u}} + \nabla \cdot \boldsymbol{q} - H = 0, \tag{6.1}$$

where ρ is the mass density, w is the stored energy per unit mass, $\boldsymbol{T}$ is Cauchy stress, $\boldsymbol{u}$ is the displacement field, $\boldsymbol{q}$ is heat flux, and H is the rate of electromagnetic energy absorbed due to Joule-heating (a source term)

$$H = a\,(\boldsymbol{J} \cdot \boldsymbol{E})\,, \tag{6.2}$$

where $0 \leq a \leq 1$ is an absorption constant. This standard form of Joule-heating is derived in the next chapter. Thus, a material designer must seek ways by which to modify a base material in order to deliver a specified overall conductivity (for example, by employing particulate additives), while simultaneously avoiding overheating.

Our objective in this chapter is to determine the phase-wise levels of the Joule-field, denoted by $H = \boldsymbol{J} \cdot \boldsymbol{E}$, in a heterogeneous three-dimensional continuum. The outline is as follows: (1) General expressions are developed for the Joule-heating field, $H = \boldsymbol{J} \cdot \boldsymbol{E}$ in a two-phase material, (2) The expressions are then specialized to isotropic cases, (3) Two asymptotic cases are then studied: (a) high-conductivity ("superconducting") particles added to a lower relative conductivity matrix and (b) low-conductivity ("insulator") particles added to a higher relative conductivity matrix, (4) Generalizations are developed for multiphase materials, for example, coated particles embedded in

T.I. Zohdi: Electromagnetic Properties of Multiphase Dielectrics, LNACM 64, pp. 63–72.
springerlink.com

a binding matrix and (5) A summary is provided with extensions discussing numerical methods.

Remark: Another detrimental effect of large electrical fields is dielectric breakdown where, if the field is strong enough to mobilize free electrons that are present in a medium, they attain sufficiently large energies to dislodge other electrons, resulting in a large number of free electrons and positively charged ions. The dislodged electrons then repeat the procedure in a chain-like reaction manner. This effect can lead to electronic device failure.

6.2 The Joule-Fields

6.2.1 Ergodic Assumptions for the Joule-Field

Similar to the previous analysis of $\boldsymbol{D}$ and $\boldsymbol{E}$, we shall utilize the following energy/power criterion (microenergy(power) must equal the macroenergy(power))

$$\langle H \rangle_\Omega = \langle \boldsymbol{J} \cdot \boldsymbol{E} \rangle_\Omega = \langle \boldsymbol{J} \rangle_\Omega \cdot \langle \boldsymbol{E} \rangle_\Omega, \tag{6.3}$$

which is an ergodicity condition (see Kröner [52], Torquato [100] and Hill [43]). Equation 6.3 is developed by first splitting the current and electric fields into a purely fluctuating part (zero mean) $\boldsymbol{J} = \langle \boldsymbol{J} \rangle_\Omega + \tilde{\boldsymbol{J}}$, where $\langle \tilde{\boldsymbol{J}} \rangle_\Omega = \boldsymbol{0}$ and $\boldsymbol{E} = \langle \boldsymbol{E} \rangle_\Omega + \tilde{\boldsymbol{E}}$ where $\langle \tilde{\boldsymbol{E}} \rangle_\Omega = \boldsymbol{0}$. By direct expansion of the system energy/power, we obtain

$$\langle (\langle \boldsymbol{J} \rangle_\Omega + \tilde{\boldsymbol{J}}) \cdot (\langle \boldsymbol{E} \rangle_\Omega + \tilde{\boldsymbol{E}}) \rangle_\Omega = \langle \boldsymbol{J} \rangle_\Omega \cdot \langle \boldsymbol{E} \rangle_\Omega + \langle \tilde{\boldsymbol{J}} \cdot \tilde{\boldsymbol{E}} \rangle_\Omega \tag{6.4}$$

since $\langle \tilde{\boldsymbol{J}} \rangle_\Omega = \boldsymbol{0}$ and $\langle \tilde{\boldsymbol{E}} \rangle_\Omega = \boldsymbol{0}$. The ergodicity assumption is that $\langle \tilde{\boldsymbol{J}} \cdot \tilde{\boldsymbol{E}} \rangle_\Omega \to 0$, as the volume, $|\Omega| \to \infty$ (relative to the inherent length-scales in the microstructure). The implication is that, as the sample becomes infinitely large, $\tilde{\boldsymbol{J}} \cdot \tilde{\boldsymbol{E}}$ is purely fluctuating and hence $\langle \tilde{\boldsymbol{J}} \cdot \tilde{\boldsymbol{E}} \rangle_\Omega = 0$. In other words, the product of two purely fluctuating random fields is also purely fluctuating.

Remark: Usually, a numerical analyst will apply uniform loading on a large sample, with the understanding that this idealization reproduces what a representative volume element (which is much smaller than the structural component of intended use) would experience within a structure. Uniform loading is an idealization and will be present within a vanishingly small microstructure relative to a finite-sized engineering (macro)structure. The macro/micro criterion is commonly used to help determine the appropriate heterogeneous material sample size for numerical effective property calculations (Zohdi and Wriggers [128]).

6.2.2 Decomposition of the Joule-Field

The Joule-fields can be written as

$$\langle H\rangle_{\Omega_i} \stackrel{\text{def}}{=} \langle \boldsymbol{J}\rangle_{\Omega_i} \cdot \langle \boldsymbol{E}\rangle_{\Omega_i} = \boldsymbol{\sigma}_i^{-1} \cdot \langle \boldsymbol{J}\rangle_{\Omega_i} \cdot \langle \boldsymbol{J}\rangle_{\Omega_i} = \boldsymbol{\sigma}_i \cdot \langle \boldsymbol{E}\rangle_{\Omega_i} \cdot \langle \boldsymbol{E}\rangle_{\Omega_i}, \quad (6.5)$$

or explicitly in terms of the overall fields, utilizing the concentration tensors introduced in the previous chapter,

$$\langle H\rangle_{\Omega_i} \stackrel{\text{def}}{=} \langle \boldsymbol{J}\rangle_{\Omega_i} \cdot \langle \boldsymbol{E}\rangle_{\Omega_i} = (\boldsymbol{C}_{J,i} \cdot \langle \boldsymbol{J}\rangle_{\Omega}) \cdot (\boldsymbol{C}_{E_i} \cdot \langle \boldsymbol{E}\rangle_{\Omega}), \quad (6.6)$$

or purely the overall current field

$$\langle H\rangle_{\Omega_i} \stackrel{\text{def}}{=} \langle \boldsymbol{J}\rangle_{\Omega_i} \cdot \langle \boldsymbol{E}\rangle_{\Omega_i} = \boldsymbol{\sigma}_i^{-1} \cdot \langle \boldsymbol{J}\rangle_{\Omega_i} \cdot \langle \boldsymbol{J}\rangle_{\Omega_i} = \boldsymbol{\sigma}_i^{-1} \cdot (\boldsymbol{C}_{J,i} \cdot \langle \boldsymbol{J}\rangle_{\Omega}) \cdot (\boldsymbol{C}_{J_i} \cdot \langle \boldsymbol{J}\rangle_{\Omega}), \quad (6.7)$$

or purely the overall electric field

$$\langle H\rangle_{\Omega_i} \stackrel{\text{def}}{=} \langle \boldsymbol{J}\rangle_{\Omega_i} \cdot \langle \boldsymbol{E}\rangle_{\Omega_i} = \boldsymbol{\sigma}_i \cdot \langle \boldsymbol{E}\rangle_{\Omega_i} \cdot \langle \boldsymbol{E}\rangle_{\Omega_i} = \boldsymbol{\sigma}_i \cdot (\boldsymbol{C}_{E,i} \cdot \langle \boldsymbol{E}\rangle_{\Omega}) \cdot (\boldsymbol{C}_{E_i} \cdot \langle \boldsymbol{E}\rangle_{\Omega}). \quad (6.8)$$

6.3 Limits on Fields

One can adopt various criteria for limits on the fields in each phase, for example,

- **Criterion** #1: Limits on the electrical fields: $||\langle \boldsymbol{E}\rangle_{\Omega_i}|| = ||\boldsymbol{C}_{E,i} \cdot \langle \boldsymbol{E}\rangle_{\Omega}|| \leq E_{i,crit}$, thus

$$||\langle \boldsymbol{E}\rangle_{\Omega_i}|| \leq ||\boldsymbol{C}_{E,i}||||\langle \boldsymbol{E}\rangle_{\Omega}|| \leq E_{i,crit} \Rightarrow ||\langle \boldsymbol{E}\rangle_{\Omega}|| \leq \frac{E_{i,crit}}{||\boldsymbol{C}_{E,i}||}, \quad (6.9)$$

  and in the case of isotropy ($\boldsymbol{C}_{E,i} = C_{E,i}\boldsymbol{1}$)

$$||\langle \boldsymbol{E}\rangle_{\Omega}|| \leq \frac{E_{i,crit}}{C_{E,i}}. \quad (6.10)$$

- **Criterion** #2: Limits on the current fields: $||\langle \boldsymbol{J}\rangle_{\Omega_i}|| = ||\boldsymbol{C}_{J,i} \cdot \langle \boldsymbol{J}\rangle_{\Omega}|| \leq J_{i,crit}$, thus

$$||\langle \boldsymbol{J}\rangle_{\Omega_i}|| \leq ||\boldsymbol{C}_{J,i}||||\langle \boldsymbol{J}\rangle_{\Omega}|| \leq J_{i,crit} \Rightarrow ||\langle \boldsymbol{J}\rangle_{\Omega}|| \leq \frac{J_{i,crit}}{||\boldsymbol{C}_{J,i}||}, \quad (6.11)$$

  and in the case of isotropy ($\boldsymbol{C}_{J,i} = C_{J,i}\boldsymbol{1}$)

$$||\langle \boldsymbol{J}\rangle_{\Omega}|| \leq \frac{J_{i,crit}}{C_{J,i}}. \quad (6.12)$$

- **Criterion** #3: Limits on the Joule-fields: $\langle H\rangle_{\Omega_i} = \langle \boldsymbol{J}\rangle_{\Omega_i} \cdot \langle \boldsymbol{E}\rangle_{\Omega_i} = (\boldsymbol{C}_{J,i} \cdot \langle \boldsymbol{J}\rangle_{\Omega}) \cdot (\boldsymbol{C}_{E_i} \cdot \langle \boldsymbol{E}\rangle_{\Omega}) \leq J_{i,crit}E_{i,crit}$, thus $(\langle H\rangle_{\Omega} \stackrel{\text{def}}{=} \langle \boldsymbol{J}\rangle_{\Omega} \cdot \langle \boldsymbol{E}\rangle_{\Omega})$,

$$\langle H\rangle_{\Omega_i} = \langle \boldsymbol{J}\rangle_{\Omega_i} \cdot \langle \boldsymbol{E}\rangle_{\Omega_i} \leq ||\boldsymbol{C}_{E,i}||||\boldsymbol{C}_{J,i}||\langle H\rangle_{\Omega} \leq J_{i,crit}E_{i,crit} \Rightarrow \langle H\rangle_{\Omega} \leq \frac{J_{i,crit}E_{i,crit}}{||\boldsymbol{C}_{E,i}||||\boldsymbol{C}_{J,i}||}, \tag{6.13}$$

which, in the case of isotropy, becomes

$$\langle H\rangle_{\Omega} \leq \frac{J_{i,crit}E_{i,crit}}{C_{E,i}C_{J,i}}, \tag{6.14}$$

or, alternatively, in terms of currents (assuming $||\boldsymbol{\sigma}_i^{-1}||J_{i,crit} = E_{i,crit}$), which collapses to Equation 6.11

$$\langle H\rangle_{\Omega_i} = \boldsymbol{\sigma}_i^{-1} \cdot \langle \boldsymbol{J}\rangle_{\Omega_i} \cdot \langle \boldsymbol{J}\rangle_{\Omega_i} \leq ||\boldsymbol{\sigma}_i^{-1}||||\boldsymbol{C}_{J,i}||||\boldsymbol{C}_{J,i}||||\langle \boldsymbol{J}\rangle_{\Omega}||^2 \Rightarrow ||\langle \boldsymbol{J}\rangle_{\Omega}|| \leq \frac{J_{i,crit}}{||\boldsymbol{C}_{J,i}||}, \tag{6.15}$$

which, in the case of isotropy, collapses to Equation 6.12, or, alternatively, in terms of electrical fields (which collapses to Equation 6.9 (assuming $J_{i,crit} = ||\boldsymbol{\sigma}_i||E_{i,crit}$))

$$\langle H\rangle_{\Omega_i} = \boldsymbol{\sigma}_i \cdot \langle \boldsymbol{E}\rangle_{\Omega_i} \cdot \langle \boldsymbol{E}\rangle_{\Omega_i} \leq ||\boldsymbol{\sigma}||||\boldsymbol{C}_{E,i}||||\boldsymbol{C}_{E,i}||||\langle \boldsymbol{E}\rangle_{\Omega}||^2 \Rightarrow ||\langle \boldsymbol{E}\rangle_{\Omega}|| \leq \frac{E_{i,crit}}{||\boldsymbol{C}_{E,i}||}, \tag{6.16}$$

which in the case of isotropy, this collapses to Equation 6.10.

We remark that Criterion #3 results from the product of Criterion #1 and Criterion #2. For either of the three criteria, because the concentration functions depend on $\boldsymbol{\sigma}^*$ and $\boldsymbol{\sigma}^* = \mathcal{F}(\boldsymbol{\sigma}_1, \boldsymbol{\sigma}_2, v_2, microstructure)$, we need to employ estimates for $\boldsymbol{\sigma}^*$, for example, using effective property bounding principles.

6.4 Utilization of Effective Property Bounds

There are a number of methods to estimate the overall macroscopic properties of materials consisting of a matrix, containing a uniform distribution of particles, in terms of the individual phase volume fractions and properties. For an in-depth review of such techniques and the general theory of random heterogeneous media, see Torquato [100] for general interdisciplinary discussions, Jikov et al. [47] for more mathematical aspects, Hashin [39], Mura [72], Nemat–Nasser and Hori [74] for solid-mechanics inclined accounts of the subject, Sevostianov and Kachanov [95] for analyses of cracked media and Zohdi and Wriggers [128] for computational aspects.

As mentioned previously, a widely-used set of estimates (in fact, bounds) for isotropic materials (such as an isotropic matrix with randomly dispersed isotropic particles) are the Hashin–Shtrikman bounds (Hashin and Shtrikman [42])

$$\underbrace{\sigma_1 + \frac{v_2}{\frac{1}{\sigma_2-\sigma_1}+\frac{1-v_2}{3\sigma_1}}}_{\sigma^{*,-}} \leq \sigma^* \leq \underbrace{\sigma_2 + \frac{1-v_2}{\frac{1}{\sigma_1-\sigma_2}+\frac{v_2}{3\sigma_2}}}_{\sigma^{*,+}}, \tag{6.17}$$

where $\sigma_2 \geq \sigma_1$, v_2 is the volume fraction of phase with the higher σ value ("phase 2" (σ_2) in the former expression) for the conductivity-mismatch. Such bounds are the tightest for isotropic effective responses, with isotropic two phase microstructures, where only the volume fractions and phase contrasts of the constituents are known. The lower bound is typically more accurate for microstructures where high conductivity particles are surrounded by a low conductivity matrix, while the upper bound is more accurate for a high conductivity matrix surrounding low conductivity particles. This can be explained qualitatively in the following manner. Consider a heterogeneous material with 50 % low conductivity material and 50 % high conductivity material. Now, consider a case (Case 1) where the matrix material is made of the low conductivity material and the particles are comprised of the high conductivity material and the opposite case (Case 2) where the matrix material is made of the high conductivity material and the particles are comprised of the low conductivity material. The bounds in both cases are the same, however, the actual overall conductivity in Case 2 is clearly higher than the overall conductivity in Case 1. Clearly, Case 2 is more closely approximated by the upper bound and Case 1 is closer to the lower bound. In summary, since the true effective property lies between the upper and lower bounds, one can construct the following approximation:

$$\sigma^* \approx \phi\sigma^{*,+} + (1-\phi)\sigma^{*,-}, \tag{6.18}$$

where $0 \leq \phi \leq 1$. ϕ *is an unknown function of the microstructure.* However, the general trends are that, for cases where the upper bound is more accurate, $\phi > \frac{1}{2}$, while for cases when the lower bound is more accurate, $\phi < \frac{1}{2}$. We now investigate these cases further.

Remark: There is a direct analogy to a the elastic properties of a material comprised of a stiff (high conductivity) matrix with embedded soft (low conductivity) inclusions versus a material comprised of a soft (low conductivity) matrix containing hard (high conductivity) inclusions.

6.5 Asymptotic Limits: Superconductors and Insulators

6.5.1 Case 1: High-Conductivity ("Superconducting") Particles

For the case of high-conductivity particles (phase 2) in a lower-conductivity matrix (phase 1), we have

$$1 << \frac{\sigma_2}{\sigma_1} \stackrel{\text{def}}{=} \alpha. \tag{6.19}$$

Inserting this expression into the Hashin–Shtrikman bounds and taking the limit as $\alpha \to \infty$ yields (σ_2 tending to infinity)

$$\sigma_1 \left(\frac{1 + 2v_2}{1 - v_2} \right) \stackrel{\text{def}}{=} \sigma_1 \zeta \leq \sigma^* \leq \infty, \tag{6.20}$$

where the lower (Hashin–Shtrikman) bound is more accurate ($\phi \to 0$). Correspondingly, for the concentration tensors for phase 1 (assuming isotropy) [1]

$$C_{E,1} = \frac{1}{1 - v_2} \quad \text{and} \quad C_{J,1} = \frac{1}{\zeta(1 - v_2)} = \frac{1}{1 + 2v_2}, \tag{6.22}$$

and for phase 2 (particle)

$$C_{E,2} = 0 \quad \text{and} \quad C_{J,2} = \frac{1}{v_2}\left(1 - \frac{1}{\zeta}\right) = \frac{3}{1 + 2v_2}. \tag{6.23}$$

The expressions are appropriate for small v_2 (superconducting particles in a binding matrix). For the various limiting criteria, we have

- **Criterion** #1: In order to *not* exceed the electrical field strength for phase 1 (matrix),

$$||\langle \boldsymbol{E} \rangle_\Omega|| \leq \frac{E_{1,crit}}{C_{E,1}} = E_{1,crit}(1 - v_2), \tag{6.24}$$

  and for phase 2 (particle),

$$||\langle \boldsymbol{E} \rangle_\Omega|| \leq \frac{E_{2,crit}}{C_{E,2}} = \infty. \tag{6.25}$$

- **Criterion** #2: In order for the loading *not* to exceed the current field strength for phase 1 (matrix),

$$||\langle \boldsymbol{J} \rangle_\Omega|| \leq \frac{J_{1,crit}}{C_{J,1}} = J_{1,crit}(1 + 2v_2), \tag{6.26}$$

[1] These expressions are consistent with the identities

$$v_1 C_{E,1} + v_2 C_{E,2} = 1 \quad \text{and} \quad v_1 C_{J,1} + v_2 C_{J,2} = 1. \tag{6.21}$$

and for phase 2 (particle)

$$||\langle \boldsymbol{J} \rangle_{\Omega}|| \leq \frac{J_{2,crit}}{C_{J,2}} = J_{2,crit}\frac{(1+2v_2)}{3}. \tag{6.27}$$

- **Criterion** #3: In order for the loading *not* to exceed the Joule-field strength for phase 1 (matrix),

$$\langle H \rangle_{\Omega_1} \leq \frac{\langle H \rangle_{\Omega}}{(1-v_2)(1+2v_2)} \leq J_{1,crit}E_{1,crit} \Rightarrow \langle H \rangle_{\Omega} \leq J_{1,crit}E_{1,crit}(1-v_2)(1+2v_2), \tag{6.28}$$

while for phase 2 (particle superconductor, no Joule-field),

$$\langle H \rangle_{\Omega_2} = 0. \tag{6.29}$$

As noted earlier, Criterion #3 results from the product of Criterion #1 and Criterion #2.

Remark: As $v_2 \to 0$ (no particle (phase 2) material), the expressions collapse to restrictions on the pure matrix (here, phase 1) material.

6.5.2 Case 2: Low-Conductivity ("Insulator") Particles

For the case of low-conductivity particles (phase 1) in a higher-conductivity matrix (phase 2), we have

$$1 >> \frac{\sigma_1}{\sigma_2} \stackrel{\text{def}}{=} \gamma. \tag{6.30}$$

Inserting this expression into the Hashin–Shtrikman bounds and taking the limit as $\gamma \to 0$ (σ_1 tending to zero) yields

$$0 \leq \sigma^* \leq \sigma_2 \left(\frac{2v_2}{3-v_2}\right) \stackrel{\text{def}}{=} \sigma_2\lambda, \tag{6.31}$$

where the upper (Hashin–Shtrikman) bound is more accurate ($\phi \to 1$). Correspondingly, for the concentration tensors (as $\gamma \to 0$), for phase 1 (particle),

$$C_{E,1} = \frac{1-\lambda}{1-v_2} = \frac{3}{3-v_2} \quad \text{and} \quad C_{J,1} = 0, \tag{6.32}$$

and for phase 2 (matrix),

$$C_{E,2} = \frac{\lambda}{v_2} = \frac{2}{3-v_2} \quad \text{and} \quad C_{J,2} = \frac{1}{v_2}. \tag{6.33}$$

The expressions are appropriate for large v_2 (insulating particles in a binding matrix). For the various limiting criteria, we have

- **Criterion #1:** In order to *not* exceed the electrical field strength for phase 1 (particle):

$$||\langle \boldsymbol{E}\rangle_{\Omega}|| \leq \frac{E_{1,crit}}{C_{E,1}} = E_{1,crit}\frac{3-v_2}{3}, \tag{6.34}$$

and for phase 2 (matrix)

$$||\langle \boldsymbol{E}\rangle_{\Omega}|| \leq \frac{E_{2,crit}}{C_{E,2}} = E_{2,crit}\frac{3-v_2}{2}. \tag{6.35}$$

- **Criterion #2:** In order for the loading *not* to exceed the current field strength for phase 1 (particle)

$$||\langle \boldsymbol{J}\rangle_{\Omega}|| \leq \frac{J_{1,crit}}{C_{J,1}} = \infty, \tag{6.36}$$

and for phase 2 (matrix)

$$||\langle \boldsymbol{J}\rangle_{\Omega}|| \leq \frac{J_{2,crit}}{C_{J,2}} = J_{2,crit}v_2. \tag{6.37}$$

- **Criterion #3:** In order for the loading *not* to exceed the Joule-field strength for phase 1 (particle insulator, no Joule-field)

$$\langle H\rangle_{\Omega_1} = 0, \tag{6.38}$$

and for phase 2 (matrix)

$$\langle H\rangle_{\Omega_2} \leq \frac{2\langle H\rangle_{\Omega}}{(3-v_2)v_2} \leq J_{2,crit}E_{2,crit} \Rightarrow \langle H\rangle_{\Omega} \leq J_{2,crit}E_{2,crit}\frac{(3-v_2)v_2}{2}. \tag{6.39}$$

Remark: As $v_2 \to 1$ (no particle (here, phase 1) material), the expressions collapse to restrictions on the pure matrix (here, phase 2) material.

6.6 Coated and Multiphase Materials

In some applications, (multiply) coated particles (Figure 6.1) or multiphase materials are useful in electrical applications. In either case, the load level can be written as, for N phases,

$$\begin{aligned}\langle \boldsymbol{E}\rangle_{\Omega} &= \frac{1}{|\Omega|}\Big(\int_{\Omega_1} \boldsymbol{E}\, d\Omega + \int_{\Omega_2} \boldsymbol{E}\, d\Omega + \int_{\Omega_3} \boldsymbol{E}\, d\Omega + \ldots + \int_{\Omega_N} \boldsymbol{E}\, d\Omega\Big)\\ &= v_1\langle \boldsymbol{E}\rangle_{\Omega_1} + v_2\langle \boldsymbol{E}\rangle_{\Omega_2} + v_3\langle \boldsymbol{E}\rangle_{\Omega_3} + \ldots + v_N\langle \boldsymbol{E}\rangle_{\Omega_N}\end{aligned} \tag{6.40}$$

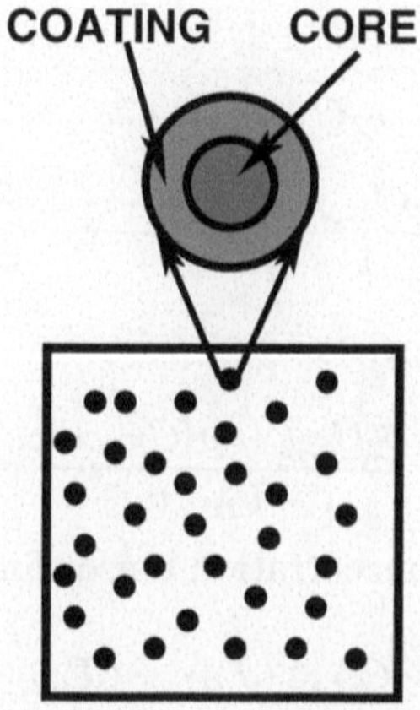

Fig. 6.1. A coated particle (Zohdi [138]).

and

$$\begin{aligned}\langle \boldsymbol{J} \rangle_{\Omega} &= \frac{1}{|\Omega|}(\int_{\Omega_1} \boldsymbol{J}\, d\Omega + \int_{\Omega_2} \boldsymbol{J}\, d\Omega + \int_{\Omega_3} \boldsymbol{J}\, d\Omega + ... + \int_{\Omega_N} \boldsymbol{J}\, d\Omega) \\ &= v_1 \langle \boldsymbol{J} \rangle_{\Omega_1} + v_2 \langle \boldsymbol{J} \rangle_{\Omega_2} + v_3 \langle \boldsymbol{J} \rangle_{\Omega_3} + ... + v_N \langle \boldsymbol{J} \rangle_{\Omega_N} \end{aligned} \tag{6.41}$$

where $(v_1+v_2+v_3+...+v_N = 1)$. Performing a set of straightforward algebraic manipulations yields

$$\begin{aligned}\langle \boldsymbol{J} \rangle_{\Omega} &= v_1 \langle \boldsymbol{J} \rangle_{\Omega_1} + v_2 \langle \boldsymbol{J} \rangle_{\Omega_2} + v_3 \langle \boldsymbol{J} \rangle_{\Omega_3} + ... + v_N \langle \boldsymbol{J} \rangle_{\Omega_N} \\ &= v_1 \boldsymbol{\sigma}_1 \cdot \langle \boldsymbol{E} \rangle_{\Omega_1} + v_2 \boldsymbol{\sigma}_2 \cdot \langle \boldsymbol{E} \rangle_{\Omega_2} + v_3 \boldsymbol{\sigma}_3 \cdot \langle \boldsymbol{E} \rangle_{\Omega_3} + ... + v_N \boldsymbol{\sigma}_N \cdot \langle \boldsymbol{E} \rangle_{\Omega_N} \\ &= \boldsymbol{\sigma}_1 \cdot (\langle \boldsymbol{E} \rangle_{\Omega} - v_2 \langle \boldsymbol{E} \rangle_{\Omega_2} - v_3 \langle \boldsymbol{E} \rangle_{\Omega_3} - ... - v_N \langle \boldsymbol{E} \rangle_{\Omega_N}) \\ &+ v_2 \boldsymbol{\sigma}_2 \cdot \langle \boldsymbol{E} \rangle_{\Omega_2} + v_3 \boldsymbol{\sigma}_3 \cdot \langle \boldsymbol{E} \rangle_{\Omega_3} + ... + v_N \boldsymbol{\sigma}_N \cdot \langle \boldsymbol{E} \rangle_{\Omega_N} \\ &= \underbrace{(\boldsymbol{\sigma}_1 + v_2(\boldsymbol{\sigma}_2 - \boldsymbol{\sigma}_1) \cdot \boldsymbol{C}_{E,2} + v_3(\boldsymbol{\sigma}_3 - \boldsymbol{\sigma}_1) \cdot \boldsymbol{C}_{E,3} + ... + v_N(\boldsymbol{\sigma}_N - \boldsymbol{\sigma}_1) \cdot \boldsymbol{C}_{E,N})}_{\boldsymbol{\sigma}^*} \cdot \langle \boldsymbol{E} \rangle_{\Omega}, \end{aligned} \tag{6.42}$$

where $\boldsymbol{C}_{E,i} \cdot \langle \boldsymbol{E} \rangle_{\Omega} = \langle \boldsymbol{E} \rangle_{\Omega_i}$, for $i = 2, 3, ..., N$, given explicitly by

$$\boldsymbol{C}_{E,i} = \frac{(\boldsymbol{\sigma}_i - \boldsymbol{\sigma}_1)^{-1}}{v_i} \cdot ((\boldsymbol{\sigma}^* - \boldsymbol{\sigma}_1) - \sum_{j \neq 1, j \neq i} v_i(\boldsymbol{\sigma}_i - \boldsymbol{\sigma}_1) \cdot \boldsymbol{C}_{E,j}), \tag{6.43}$$

which leads to a coupled set of equations for the $\boldsymbol{C}_{E,i}$. For phase 1, since

$$\underbrace{v_1 \boldsymbol{C}_{E,1}}_{\text{phase}-1 \text{ contribution}} + \underbrace{v_2 \boldsymbol{C}_{E,2}}_{\text{phase}-2 \text{ contribution}} + \underbrace{v_3 \boldsymbol{C}_{E,3}}_{\text{phase}-3 \text{ contribution}} + ... + \underbrace{v_N \boldsymbol{C}_{E,N}}_{\text{phase}-\text{N contribution}} = \boldsymbol{1}, \tag{6.44}$$

then

$$\begin{aligned} \boldsymbol{C}_{E,1} &\stackrel{\text{def}}{=} \frac{1}{v_1}(\mathbf{1} - v_2\boldsymbol{C}_{E,2} - v_3\boldsymbol{C}_{E,3} - ... + -v_N\boldsymbol{C}_{E,N}) \\ &= \frac{\mathbf{1} - v_2\boldsymbol{C}_{E,2} - v_3\boldsymbol{C}_{E,3} - ... - v_N\boldsymbol{C}_{E,N}}{1 - v_2 - v_3 - ... - v_N} \end{aligned} \tag{6.45}$$

where, in the case of isotropy,

$$C_{E,1} = \frac{1 - v_2 C_{E,2} - v_3 C_{E,3} - ... - v_N C_{E,N}}{1 - v_2 - v_3 - ... - v_N}. \tag{6.46}$$

For the current field, the concentration tensor for phase i

$$\boldsymbol{C}_{J,i} \cdot \langle \boldsymbol{J} \rangle_{\Omega} = \langle \boldsymbol{J} \rangle_{\Omega_i} \tag{6.47}$$

where $\boldsymbol{C}_{J,i} = \boldsymbol{\sigma}_i \cdot \boldsymbol{C}_{E,i} \cdot \boldsymbol{\sigma}^{*-1}$. The Criteria #1 – #3 are still governed by Equations 6.9-6.16, with virtually no modification. *Once either the set of $\boldsymbol{C}_{E,i}$ or $\boldsymbol{\sigma}^*$ are known, the other can be determined.* However, *sharp* effective property bounds for N-phase composites are lacking in most cases, and one should probably resort to numerical discretization methods.

6.7 Summary and Discussion

In principle, the approach introduced here can be used with other bounding or estimation techniques. To the knowledge of the author, Joule-field amplification estimates have not been previously investigated, although various bounds and estimates for other phase-wise fields can be found, for example, in the book of Kreher and Pompe [51]. The more general problem of estimating higher moments of the fields, in particular the second moments and the full distribution of the phase fields, can be found in Bergman [7], Cule and Torquato [12], as well as in a series of papers by Lipton, who explored local fields in-depth (Lipton [59–62]). More detailed information, for example, localized effects in the matrix ligaments between particles ("hot spots"), need to be generated numerically by solving a boundary-value problem posed over a statistically representative volume element (RVE) sample of heterogeneous media. In particular, since time-transient effects lead to coupling of electrical and magnetic fields, the only viable approach is to employ direct numerical techniques to solve for Maxwell's equations. Additionally, if the local material properties are thermally-sensitive, and Joule-heating is significant, then the first law of thermodynamics must also be solved, simultaneously, in order to determine the temperature. Numerical techniques for the solution of coupled boundary value problems posed over heterogeneous electromagnetic media are developed later in the monograph.

7

Some Basic Principles of Continuum Mechanics

In order to properly consider multifield coupling effects, we will need to draw on some of the tools of classical continuum mechanics.

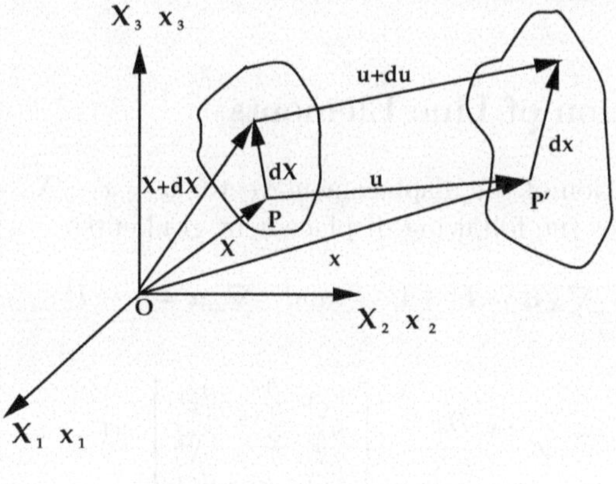

Fig. 7.1. A deforming body.

7.1 Motion and Deformation

The term deformation refers to a change in the shape of the continuum between a reference configuration and current configuration. In the reference configuration, a representative particle of the continuum occupies a point $\mathbf{p}$ in space and has the position vector

T.I. Zohdi: Electromagnetic Properties of Multiphase Dielectrics, LNACM 64, pp. 73–88.
springerlink.com

$$\boldsymbol{X} = X_1\boldsymbol{e}_1 + X_2\boldsymbol{e}_2 + X_3\boldsymbol{e}_3$$

where $\boldsymbol{e}_1, \boldsymbol{e}_2, \boldsymbol{e}_3$ is a Cartesian reference triad, and X_1, X_2, X_3 (with center **O**) can be thought of as labels for a point. Sometimes, the coordinates or labels (X_1, X_2, X_3, t) are called the referential coordinates. In the current configuration, the particle originally located at point **P** is located at point **P**$'$, and can also be expressed in terms of another position vector $\boldsymbol{x}$, with the coordinates (x_1, x_2, x_3, t). These are called the current coordinates. It is obvious with this arrangement that the displacement is $\boldsymbol{u} = \boldsymbol{x} - \boldsymbol{X}$ for a point originally at $\boldsymbol{X}$ and with final coordinates $\boldsymbol{x}$.

When a continuum undergoes deformation (or flow), its points move along various paths in space. This motion may be expressed by

$$\boldsymbol{x}(X_1, X_2, X_3, t) = \boldsymbol{u}(X_1, X_2, X_3, t) + \boldsymbol{X}(X_1, X_2, X_3, t)\,,$$

which gives the present location of a point at time t, written in terms of the labels X_1, X_2, X_3. The previous position vector ($\boldsymbol{x}$) may be interpreted as a mapping of the initial configuration onto the current configuration. In classical approaches, it is assumed that such a mapping is one-to-one and continuous, with continuous partial derivatives to whatever order is required. The description of motion or deformation expressed previously is known as the Lagrangian formulation. Alternatively, if the independent variables are the coordinates $\boldsymbol{x}$ and t, then $\boldsymbol{x}(x_1, x_2, x_3, t) = \boldsymbol{u}(x_1, x_2, x_3, t) + \boldsymbol{X}(x_1, x_2, x_3, t)$, and the formulation is denoted as Eulerian.

7.2 Deformation of Line Elements

Partial differentiation of the displacement vector $\boldsymbol{u} = \boldsymbol{x} - \boldsymbol{X}$, with respect to $\boldsymbol{x}$ and $\boldsymbol{X}$, produces the following displacement gradients:

$$\nabla_X \boldsymbol{u} = \boldsymbol{F} - \boldsymbol{1} \qquad \text{and} \qquad \nabla_x \boldsymbol{u} = \boldsymbol{1} - \overline{\boldsymbol{F}} \tag{7.1}$$

where

$$\nabla_X \boldsymbol{x} \stackrel{\text{def}}{=} \frac{\partial \boldsymbol{x}}{\partial \boldsymbol{X}} = \begin{bmatrix} \frac{\partial x_1}{\partial X_1} & \frac{\partial x_1}{\partial X_2} & \frac{\partial x_1}{\partial X_3} \\ \frac{\partial x_2}{\partial X_1} & \frac{\partial x_2}{\partial X_2} & \frac{\partial x_2}{\partial X_3} \\ \frac{\partial x_3}{\partial X_1} & \frac{\partial x_3}{\partial X_2} & \frac{\partial x_3}{\partial X_3} \end{bmatrix} \stackrel{\text{def}}{=} \boldsymbol{F}, \tag{7.2}$$

and

$$\nabla_x \boldsymbol{X} \stackrel{\text{def}}{=} \frac{\partial \boldsymbol{X}}{\partial \boldsymbol{x}} = \overline{\boldsymbol{F}} \tag{7.3}$$

with the components $F_{ik} = x_{i,k}$ and $\overline{F}_{ik} = X_{i,k}$. $\boldsymbol{F}$ is known as the material deformation gradient, and $\overline{\boldsymbol{F}}$ is known as the spatial deformation gradient.

Remark: It should be clear that $d\boldsymbol{x}$ can be reinterpreted as the result of a mapping $\boldsymbol{F} \cdot d\boldsymbol{X} \rightarrow d\boldsymbol{x}$, or a change in configuration (reference to current), while $\overline{\boldsymbol{F}} \cdot d\boldsymbol{x} \rightarrow d\boldsymbol{X}$ maps the current to the reference system. For the deformations to be invertible and physically realizable, $\overline{\boldsymbol{F}} \cdot (\boldsymbol{F} \cdot d\boldsymbol{X}) = d\boldsymbol{X}$ and $\boldsymbol{F} \cdot (\overline{\boldsymbol{F}} \cdot d\boldsymbol{x}) = d\boldsymbol{x}$. We note that $(det\overline{\boldsymbol{F}})(det\boldsymbol{F}) = 1$ and have the following obvious relation $\frac{\partial \boldsymbol{X}}{\partial \boldsymbol{x}} \cdot \frac{\partial \boldsymbol{x}}{\partial \boldsymbol{X}} = \overline{\boldsymbol{F}} \cdot \boldsymbol{F} = \boldsymbol{1}$. It should be clear that $\overline{\boldsymbol{F}} = \boldsymbol{F}^{-1}$.

7.3 The Jacobian of the Deformation Gradient

The Jacobian of the deformation gradient, $\boldsymbol{F}$, is defined as

$$J \stackrel{\text{def}}{=} det\boldsymbol{F} = \begin{vmatrix} \frac{\partial x_1}{\partial X_1} & \frac{\partial x_1}{\partial X_2} & \frac{\partial x_1}{\partial X_3} \\ \frac{\partial x_2}{\partial X_1} & \frac{\partial x_2}{\partial X_2} & \frac{\partial x_2}{\partial X_3} \\ \frac{\partial x_3}{\partial X_1} & \frac{\partial x_3}{\partial X_2} & \frac{\partial x_3}{\partial X_3} \end{vmatrix}. \tag{7.4}$$

To interpret the Jacobian in a physical way, consider a reference differential volume which is given by $dS^3 = d\omega_o$, where $d\boldsymbol{X}^{(1)} = dS\,\boldsymbol{e}_1$, $d\boldsymbol{X}^{(2)} = dS\,\boldsymbol{e}_2$ and $d\boldsymbol{X}^{(3)} = dS\,\boldsymbol{e}_3$. The current differential element is described by $d\boldsymbol{x}^{(1)} = \frac{\partial \boldsymbol{x}_k}{\partial X_1} dS\,\boldsymbol{e}_k$, $d\boldsymbol{x}^{(2)} = \frac{\partial \boldsymbol{x}_k}{\partial X_2} dS\,\boldsymbol{e}_k$ and $d\boldsymbol{x}^{(3)} = \frac{\partial \boldsymbol{x}_k}{\partial X_3} dS\,\boldsymbol{e}_k$, where $\boldsymbol{e}$ is a unit vector, and

$$\underbrace{d\boldsymbol{x}^{(1)} \cdot (d\boldsymbol{x}^{(2)} \times d\boldsymbol{x}^{(3)})}_{\stackrel{\text{def}}{=} d\omega} = \begin{vmatrix} dx_1^{(1)} & dx_2^{(1)} & dx_3^{(1)} \\ dx_1^{(2)} & dx_2^{(2)} & dx_3^{(2)} \\ dx_1^{(3)} & dx_2^{(3)} & dx_3^{(3)} \end{vmatrix} = \begin{vmatrix} \frac{\partial x_1}{\partial X_1} & \frac{\partial x_2}{\partial X_1} & \frac{\partial x_3}{\partial X_1} \\ \frac{\partial x_1}{\partial X_2} & \frac{\partial x_2}{\partial X_2} & \frac{\partial x_3}{\partial X_2} \\ \frac{\partial x_1}{\partial X_3} & \frac{\partial x_2}{\partial X_3} & \frac{\partial x_3}{\partial X_3} \end{vmatrix} dS^3. \tag{7.5}$$

Therefore, $d\omega = J\,d\omega_0$. Thus, the Jacobian of the deformation gradient must remain positive definite, otherwise we obtain physically impossible "negative" volumes.

7.4 Equilibrium/Kinetics of Solid Continua

We start with the following postulated balance law for an arbitrary part ω around a point P with boundary $\partial\omega$ of a body Ω,

$$\underbrace{\int_{\partial\omega} \boldsymbol{t}\,dA}_{\text{surface forces}} + \underbrace{\int_{\omega} \boldsymbol{f}\,d\omega}_{\text{body forces}} = \underbrace{\frac{d}{dt}\int_{\omega} \rho\dot{\boldsymbol{u}}\,d\omega}_{\text{inertial forces}}, \tag{7.6}$$

where ρ is the material density, $\boldsymbol{b}$ is the body force per unit mass ($\boldsymbol{f} = \rho\boldsymbol{b}$) and $\dot{\boldsymbol{u}}$ is the time derivative of the displacement.[1]

When the actual molecular structure is considered on a submicroscopic scale, the force densities, $\boldsymbol{t}$, which we commonly refer to as "surface forces," are taken to involve short-range intermolecular forces. Tacitly, we assume that the

[1] We use the shorthand notation of $\dot{()} \stackrel{\text{def}}{=} \frac{d()}{dt}$.

effects of radiative forces, and others which do not require momentum transfer through a continuum, are negligible. This is a so-called local action postulate. As long as the volume element is large, our resultant body and surface forces may be interpreted as sums of these intermolecular forces. When we pass to larger scales, we can justifiably use the continuum concept.

7.5 Postulates on Volume and Surface Quantities

Now, consider a tetrahedron in equilibrium, as shown in Figure 7.2. From Newton's laws,

$$\boldsymbol{t}^{(n)}\Delta A^{(n)} + \boldsymbol{t}^{(-1)}\Delta A^{(1)} + \boldsymbol{t}^{(-2)}\Delta A^{(2)} + \boldsymbol{t}^{(-3)}\Delta A^{(3)} + \boldsymbol{f}\Delta V = \rho\Delta V\ddot{\boldsymbol{u}}, \quad (7.7)$$

where $\Delta A^{(n)}$ is the surface area of the face of the tetrahedron with normal $\boldsymbol{n}$, and ΔV is the tetrahedron volume. As the distance between the tetrahedron base (located at (0,0,0)) and the surface center, denoted h, goes to zero, we have $h \to 0 \Rightarrow \Delta A^{(n)} \to 0 \Rightarrow \frac{\Delta V}{\Delta A^{(n)}} \to 0$. Geometrically, we have $\frac{\Delta A^{(i)}}{\Delta A^{(n)}} = cos(x_i, x_n) \stackrel{\text{def}}{=} n_i$, and therefore $\boldsymbol{t}^{(n)} + \boldsymbol{t}^{(-1)}cos(x_1, x_n) + \boldsymbol{t}^{(-2)}cos(x_2, x_n) + \boldsymbol{t}^{(-3)}cos(x_3, x_n) = \boldsymbol{0}$.

It is clear that forces on the surface areas could be decomposed into three linearly independent components. It is convenient to express the concept of stress at a point, representing the surface forces there, pictorially represented by a cube surrounding a point. The fundamental issue that must be resolved is the characterization of these surface forces. We can represent the force density vector, the so-called "traction," on a surface by the component representation:

$$\boldsymbol{t} \stackrel{\text{def}}{=} \left\{ \begin{matrix} T_{i1} \\ T_{i2} \\ T_{i3} \end{matrix} \right\}, \quad (7.8)$$

where the second index represents the direction of the component and the first index represents the normal to corresponding coordinate plane. From this point forth, we will drop the superscript notation of $\boldsymbol{t}^{(n)}$, where it is implicit that $\boldsymbol{t} \stackrel{\text{def}}{=} \boldsymbol{t}^{(n)} = \boldsymbol{T}^T \cdot \boldsymbol{n}$, where

$$\boldsymbol{T} \stackrel{\text{def}}{=} \begin{bmatrix} T_{11} & T_{12} & T_{13} \\ T_{21} & T_{22} & T_{23} \\ T_{31} & T_{32} & T_{33} \end{bmatrix}, \quad (7.9)$$

or explicitly ($\boldsymbol{t}^{(1)} = -\boldsymbol{t}^{(-1)}$, $\boldsymbol{t}^{(2)} = -\boldsymbol{t}^{(-2)}$, $\boldsymbol{t}^{(3)} = -\boldsymbol{t}^{(-3)}$)

$$\boldsymbol{t}^{(n)} = \boldsymbol{t}^{(1)}n_1 + \boldsymbol{t}^{(2)}n_2 + \boldsymbol{t}^{(3)}n_3 = \boldsymbol{T}^T \cdot \boldsymbol{n} = \begin{bmatrix} T_{11} & T_{12} & T_{13} \\ T_{21} & T_{22} & T_{23} \\ T_{31} & T_{32} & T_{33} \end{bmatrix}^T \left\{ \begin{matrix} n_1 \\ n_2 \\ n_3 \end{matrix} \right\}, \quad (7.10)$$

where $\boldsymbol{T}$ is the so-called Cauchy stress tensor. [2]

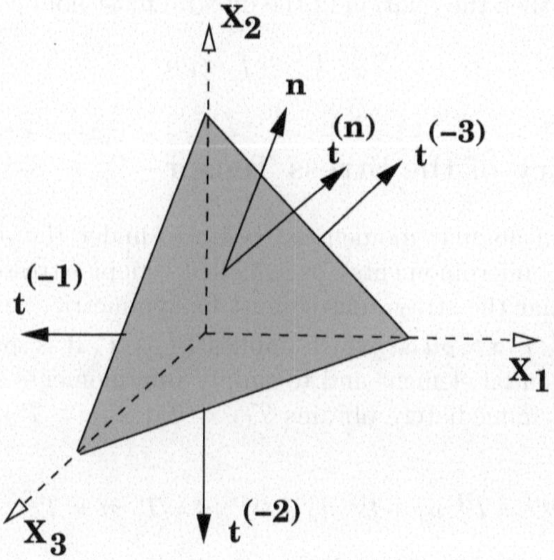

Fig. 7.2. Cauchy tetrahedron.

7.6 Balance Law Formulations

Substitution of Equation 7.10 into Equation 7.6 yields ($\omega \subset \Omega$)

$$\underbrace{\int_{\partial\omega} \boldsymbol{T}^T \cdot \boldsymbol{n}\, dA}_{\text{surface forces}} + \underbrace{\int_{\omega} \boldsymbol{f}\, d\omega}_{\text{body forces}} = \underbrace{\frac{d}{dt}\int_{\omega} \rho\dot{\boldsymbol{u}}\, d\omega}_{\text{inertial forces}}. \tag{7.11}$$

A relationship can be determined between the densities in the current and reference configurations, $\int_\omega \rho d\omega = \int_{\omega_0} \rho J d\omega_0 = \int_{\omega_0} \rho_0 d\omega_0$. Therefore, the Jacobian can also be interpreted as the ratio of material densities at a point. Since the volume is arbitrary, we can assume that $\rho J = \rho_0$ holds at every point in the body. Therefore, we may write $\frac{d}{dt}(\rho_0) = \frac{d}{dt}(\rho J) = 0$ when the system is mass conservative over time. This leads to writing the last term in

---

[2] Some authors follow the notation where the first index represents the direction of the component and the second index represents the normal to corresponding coordinate plane. This leads to $\boldsymbol{t} \stackrel{\text{def}}{=} \boldsymbol{t}^{(n)} = \boldsymbol{T}\cdot\boldsymbol{n}$. In the absence of couple stresses, a balance of angular momentum implies a symmetry of stress, $\boldsymbol{T} = \boldsymbol{T}^T$, and thus the difference in notations become immaterial.

Equation 7.6 as $\frac{d}{dt}\int_\omega \rho\dot{\boldsymbol{u}}\,d\omega = \int_{\omega_0}\frac{d(\rho J)}{dt}\dot{\boldsymbol{u}}\,d\omega_0 + \int_{\omega_0}\rho\ddot{\boldsymbol{u}}J\,d\omega_0 = \int_\omega \rho\ddot{\boldsymbol{u}}\,d\omega$. From Gauss's Divergence theorem, and an implicit assumption that $\boldsymbol{T}$ is differentiable, we have $\int_\omega \left(\nabla_x \cdot \boldsymbol{T}^T + \boldsymbol{f} - \rho\ddot{\boldsymbol{u}}\right) d\omega = \boldsymbol{0}$. If the volume is argued as being arbitrary, then the relation in the integral must hold pointwise, yielding

$$\nabla_x \cdot \boldsymbol{T}^T + \boldsymbol{f} = \rho\ddot{\boldsymbol{u}}. \tag{7.12}$$

7.7 Symmetry of the Stress Tensor

Starting with an angular momentum balance, under the assumptions that no infinitesimal "micromoments" or so-called "couple-stresses" exist, then it can be shown that the stress tensor must be symmetric, i.e., $\int_{\partial\omega} \boldsymbol{x} \times \boldsymbol{t}\,dA + \int_\omega \boldsymbol{x} \times \boldsymbol{f}\,d\omega = \frac{d}{dt}\int_\omega \boldsymbol{x} \times \rho\dot{\boldsymbol{u}}\,d\omega$ which implies $\boldsymbol{T}^T = \boldsymbol{T}$. It is somewhat easier to consider a differential element and to simply sum moments about the center. Doing this, one immediately obtains $T_{12} = T_{21}, T_{23} = T_{32}$ and $T_{13} = T_{31}$. Therefore,

$$\boldsymbol{t}^{(n)} = \boldsymbol{t}^{(1)}n_1 + \boldsymbol{t}^{(2)}n_2 + \boldsymbol{t}^{(3)}n_3 = \boldsymbol{T}\cdot\boldsymbol{n} = \boldsymbol{T}^T\cdot\boldsymbol{n}. \tag{7.13}$$

This implies that Equation 7.12 becomes

$$\nabla_x \cdot \boldsymbol{T} + \boldsymbol{f} = \rho\ddot{\boldsymbol{u}}. \tag{7.14}$$

7.8 The First Law of Thermodynamics – An Energy Balance

The interconversions of mechanical, thermal and chemical energy in a system are governed by the first law of thermodynamics, where the time rate of change of the total energy, $\mathcal{K} + \mathcal{I}$, is equal to the work rate, $\mathcal{P}_w$ and the net heat supplied, $\mathcal{H} + \mathcal{Q}$, i.e., $\frac{d}{dt}(\mathcal{K} + \mathcal{I}) = \mathcal{P}_w + \mathcal{H} + \mathcal{Q}$. Here, the kinetic energy of a subvolume of material contained in Ω, denoted ω, is $\mathcal{K} \stackrel{\text{def}}{=} \int_\omega \frac{1}{2}\rho\dot{\boldsymbol{u}}\cdot\dot{\boldsymbol{u}}\,d\omega$, the rate of work or power of external forces acting on ω is given by $\mathcal{P}_w \stackrel{\text{def}}{=} \int_\omega \rho\boldsymbol{b}\cdot\dot{\boldsymbol{u}}\,d\omega + \int_{\partial\omega} \boldsymbol{T}\cdot\boldsymbol{n}\cdot\dot{\boldsymbol{u}}\,da$, the heat flow into the volume by conduction is $\mathcal{Q} \stackrel{\text{def}}{=} -\int_{\partial\omega} \boldsymbol{q}\cdot\boldsymbol{n}\,da = -\int_\omega \nabla_x\cdot\boldsymbol{q}\,d\omega$, the heat generated due to sources, *such as chemical reactions*, is $\mathcal{H} \stackrel{\text{def}}{=} \int_\omega \rho z\,d\omega$, where z is the reaction source rate per unit mass and the stored energy is $\mathcal{I} \stackrel{\text{def}}{=} \int_\omega \rho w\,d\omega$. If we make the assumption that the mass in the system is constant, one has

$$\text{current mass} = \int_\omega \rho\,d\omega = \int_{\omega_0} \rho J\,d\omega_0 \approx \int_{\omega_0} \rho_0\,d\omega_0 = \text{original mass}, \tag{7.15}$$

which implies $\rho J = \rho_0$. Therefore, $\rho J = \rho_0 \Rightarrow \dot{\rho} J + \rho \dot{J} = 0$. Using this and the energy balance leads to

$$\begin{aligned}\frac{d\mathcal{K}}{dt} = \frac{d}{dt}\int_\omega \frac{1}{2}\rho \dot{\boldsymbol{u}} \cdot \dot{\boldsymbol{u}}\, d\omega &= \int_{\omega_0} \frac{d}{dt}\frac{1}{2}(\rho J \dot{\boldsymbol{u}} \cdot \dot{\boldsymbol{u}})\, d\omega_0 \\ &= \int_{\omega_0} (\frac{d}{dt}\rho_0)\frac{1}{2}\dot{\boldsymbol{u}} \cdot \dot{\boldsymbol{u}}\, d\omega_0 + \int_\omega \rho \frac{d}{dt}\frac{1}{2}(\dot{\boldsymbol{u}} \cdot \dot{\boldsymbol{u}})\, d\omega \\ &= \int_\omega \rho \dot{\boldsymbol{u}} \cdot \ddot{\boldsymbol{u}}\, d\omega. \end{aligned} \tag{7.16}$$

We also have

$$\frac{d\mathcal{I}}{dt} = \frac{d}{dt}\int_\omega \rho w\, d\omega = \frac{d}{dt}\int_{\omega_0} \rho J w\, d\omega_0 = \int_{\omega_0} \frac{d}{dt}(\rho_0) w\, d\omega_0 + \int_\omega \rho \dot{w}\, d\omega. \tag{7.17}$$

By using the divergence theorem, we obtain

$$\int_{\partial\omega} \boldsymbol{T} \cdot \boldsymbol{n} \cdot \dot{\boldsymbol{u}}\, da = \int_\omega \nabla_x \cdot (\boldsymbol{T} \cdot \dot{\boldsymbol{u}})\, d\omega = \int_\omega (\nabla_x \cdot \boldsymbol{T}) \cdot \dot{\boldsymbol{u}}\, d\omega + \int_\omega \boldsymbol{T} : \nabla_x \dot{\boldsymbol{u}}\, d\omega. \tag{7.18}$$

Combining the results, and enforcing a balance of linear momentum leads to

$$\begin{aligned}&\int_\omega (\rho\dot{w} + \dot{\boldsymbol{u}} \cdot (\rho\ddot{\boldsymbol{u}} - \nabla_x \cdot \boldsymbol{T} - \rho\boldsymbol{b}) - \boldsymbol{T} : \nabla_x \dot{\boldsymbol{u}} + \nabla_x \cdot \boldsymbol{q} - \rho z)\, d\omega = \\ &\int_\omega (\rho\dot{w} - \boldsymbol{T} : \nabla_x \dot{\boldsymbol{u}} + \nabla_x \cdot \boldsymbol{q} - \rho z)\, d\omega = 0.\end{aligned} \tag{7.19}$$

Since the volume ω is arbitrary, the integrand must hold locally and we have

$$\rho\dot{w} - \boldsymbol{T} : \nabla_x \dot{\boldsymbol{u}} + \nabla_x \cdot \boldsymbol{q} - \rho z = 0. \tag{7.20}$$

When dealing with multifield problems, this equation is used extensively.

7.9 Special Case: Electromagnetic Thermocoupled Problems

If the effects of deformation and stress are considered to be unimportant in the present class of problems, then

$$\rho\dot{w} + \nabla_x \cdot \boldsymbol{q} - \rho z = 0, \tag{7.21}$$

and we consider the stored thermal energy per unit mass to be

$$w = C\theta \tag{7.22}$$

and

$$\boldsymbol{q} = -\mathbb{K} \cdot \nabla_x \theta. \tag{7.23}$$

These relations lead to

$$\rho C\dot{\theta} - \nabla_x \cdot \mathbb{K} \cdot \nabla_x \theta - \rho z = 0. \tag{7.24}$$

7.10 The Poynting Vector

The energy per unit area per unit time flowing into a surface in free space is given by the Poynting vector $\boldsymbol{S} = \boldsymbol{E} \times \boldsymbol{H}$. The Poynting vector has an intimate connection to the conservation of energy. To see this, form the inner product of the magnetic field with Faraday's law, that is,

$$\boldsymbol{H} \cdot (\nabla_x \times \boldsymbol{E}) = -\boldsymbol{H} \cdot \left(\underbrace{\boldsymbol{M}^{ext} + \hat{\sigma}\boldsymbol{H}}_{\boldsymbol{M}^{tot}} + \frac{\partial \boldsymbol{B}}{\partial t} \right), \tag{7.25}$$

and the inner product of the electric field with Ampere's law,

$$\boldsymbol{E} \cdot (\nabla_x \times \boldsymbol{H}) = \boldsymbol{E} \cdot \left(\underbrace{\boldsymbol{J}^{ext} + \sigma\boldsymbol{E}}_{\stackrel{\mathrm{def}}{=} \boldsymbol{J}^{tot}} + \frac{\partial \boldsymbol{D}}{\partial t} \right). \tag{7.26}$$

Subtracting Equation 7.25 from Equation 7.26 yields

$$\underbrace{\boldsymbol{E} \cdot (\nabla_x \times \boldsymbol{H}) - \boldsymbol{H} \cdot (\nabla_x \times \boldsymbol{E})}_{-\nabla_x \cdot (\boldsymbol{E} \times \boldsymbol{H}) = -\nabla_x \cdot \boldsymbol{S}} = \boldsymbol{E} \cdot \boldsymbol{J}^{tot} + \boldsymbol{H} \cdot \boldsymbol{M}^{tot} + \underbrace{\boldsymbol{E} \cdot \frac{\partial \boldsymbol{D}}{\partial t} + \boldsymbol{H} \cdot \frac{\partial \boldsymbol{B}}{\partial t}}_{= \frac{\partial W}{\partial t}}, \tag{7.27}$$

where W is the electromagnetic energy

$$W = \frac{1}{2}(\boldsymbol{E} \cdot \boldsymbol{D} + \boldsymbol{H} \cdot \boldsymbol{B}) = \frac{1}{2}(\boldsymbol{E} \cdot \boldsymbol{\epsilon} \cdot \boldsymbol{E} + \boldsymbol{H} \cdot \boldsymbol{\mu} \cdot \boldsymbol{H}). \tag{7.28}$$

Thus,

$$\nabla_x \cdot \boldsymbol{S} + \boldsymbol{E} \cdot \boldsymbol{J}^{tot} + \boldsymbol{H} \cdot \boldsymbol{M}^{tot} + \frac{\partial W}{\partial t} = 0 \tag{7.29}$$

or

$$\frac{\partial W}{\partial t} + \nabla_x \cdot \boldsymbol{S} = -\left(\boldsymbol{E} \cdot \boldsymbol{J}^{tot} + \boldsymbol{H} \cdot \boldsymbol{M}^{tot}\right). \tag{7.30}$$

This is usually referred to as "Poynting's" theorem, and can be interpreted, *for simple material laws where Equation 7.28 holds*, as the rate of change of electromagnetic energy within a volume plus the energy flowing out through a boundary is equal to the negative of the total work done by the fields on the sources and conduction. This work is then converted into thermomechanical energy.

Remark 1: The main effect of interest is (a) the absorption of electromagnetic energy, which is converted into heat and (b) thermally-induced changes in the electromagnetic material properties. In this work, we consider the absorbed energy that is available for heating to be proportional to the energy associated with conduction, namely, from Equation 7.30, $\boldsymbol{E} \cdot \boldsymbol{J}^{tot} + \boldsymbol{H} \cdot \boldsymbol{M}^{tot}$, and account for it in the first law of thermodynamics via ρz as the rate of electromagnetic energy absorbed, namely,

$$\rho z = a \left(\boldsymbol{E} \cdot \boldsymbol{J}^{tot} + \boldsymbol{H} \cdot \boldsymbol{M}^{tot} \right), \tag{7.31}$$

where a is an absorption constant, $0 \leq a \leq 1$. This type of simple model for heating is often referred to "Joule heating."

Remark 2: One can interpret Joule heating as arising from charged particles being pulled through a medium by electromagnetic fields and giving up some of their kinetic energy when they collide with their surroundings, which manifests itself as heat.

7.11 Infinitesimal Linearly Elastic Constitutive Laws

Here, we provide a brief introduction of infinitesimal strain linearized elasticity, which we will use later, based on a more extensive discussion found in Zohdi and Wriggers [128].

7.11.1 Infinitesimal Strain Linear Elasticity

If we neglect transfer of heat and production of heat, Equation 7.20 implies $\rho \dot{w} = \boldsymbol{T} : \nabla_x \dot{\boldsymbol{u}}$, which in the infinitesimal strain linearly elastic case[3] is equivalent to $\rho \dot{w} = \boldsymbol{T} : \dot{\boldsymbol{\varepsilon}}$. From the chain rule of differentiation we have $\rho \dot{w} = \rho \frac{\partial w}{\partial \boldsymbol{\varepsilon}} : \frac{d\boldsymbol{\varepsilon}}{dt} = \boldsymbol{T} : \dot{\boldsymbol{\varepsilon}} \Rightarrow \boldsymbol{T} = \rho \frac{\partial w}{\partial \boldsymbol{\varepsilon}}$. The starting point to develop a constitutive theory is to assume a stored elastic energy function exists, a function denoted $W \stackrel{\text{def}}{=} \rho w$, which depends only on the mechanical deformation. The simplest function that fulfills $\boldsymbol{T} = \rho \frac{\partial w}{\partial \boldsymbol{\varepsilon}}$ is $W = \frac{1}{2} \boldsymbol{\varepsilon} : \mathbb{E} : \boldsymbol{\varepsilon}$. Such a function satisfies the intuitive physical requirement that, for any small strain from an undeformed state, energy must be stored in the material. Alternatively, a small strain material law can be derived from $\boldsymbol{T} = \frac{\partial W}{\partial \boldsymbol{\varepsilon}}$ and $W \approx c_0 + \boldsymbol{c}_1 : \boldsymbol{\varepsilon} + \frac{1}{2} \boldsymbol{\varepsilon} : \mathbb{E} : \boldsymbol{\varepsilon} + ...$ which implies $\boldsymbol{T} \approx \boldsymbol{c}_1 + \mathbb{E} : \boldsymbol{\varepsilon} + ...$ We are free to set $c_0 = 0$ (it is arbitrary) to have zero strain energy at zero strain, and

[3] The infinitesimal strain is defined as $\boldsymbol{\varepsilon} = \frac{1}{2}(\nabla_X \boldsymbol{u} + (\nabla_X \boldsymbol{u})^T)$

furthermore, we assume that no stresses exist in the reference state ($\boldsymbol{c}_1 = \boldsymbol{0}$). Consequently, we obtain the familiar relation

$$\boldsymbol{T} \approx \mathbb{E} : \boldsymbol{\varepsilon}. \tag{7.32}$$

This is a linear (tensorial) relation between stresses and strains. The existence of a strictly positive stored energy function at the reference configuration implies that the linear elasticity tensor, $\mathbb{E}$, must have positive eigenvalues at every point in the body. Typically, different materials are classified according to the number of independent constants in $\mathbb{E}$. A general material has 81 independent constants since it is a fourth order tensor relating nine components of stress to strain. However, the number of constants can be reduced to 36 since the stress and strain tensors are symmetric. This is easily seen from the matrix representation of $\mathbb{E}$:

$$\underbrace{\begin{Bmatrix} T_{11} \\ T_{22} \\ T_{33} \\ T_{12} \\ T_{23} \\ T_{31} \end{Bmatrix}}_{\stackrel{\text{def}}{=}\{\boldsymbol{T}\}} = \underbrace{\begin{bmatrix} E_{1111} & E_{1122} & E_{1133} & E_{1112} & E_{1123} & E_{1113} \\ E_{2211} & E_{2222} & E_{2233} & E_{2212} & E_{2223} & E_{2213} \\ E_{3311} & E_{3322} & E_{3333} & E_{3312} & E_{3323} & E_{3313} \\ E_{1211} & E_{1222} & E_{1233} & E_{1212} & E_{1223} & E_{1213} \\ E_{2311} & E_{2322} & E_{2333} & E_{2312} & E_{2323} & E_{2313} \\ E_{1311} & E_{1322} & E_{1333} & E_{1312} & E_{1323} & E_{1313} \end{bmatrix}}_{\stackrel{\text{def}}{=}[\mathbb{E}]} \underbrace{\begin{Bmatrix} \varepsilon_{11} \\ \varepsilon_{22} \\ \varepsilon_{33} \\ 2\varepsilon_{12} \\ 2\varepsilon_{23} \\ 2\varepsilon_{31} \end{Bmatrix}}_{\stackrel{\text{def}}{=}\{\boldsymbol{\varepsilon}\}}. \tag{7.33}$$

The symbol $[\cdot]$ is used to indicate standard matrix notation equivalent to a tensor form, while $\{\cdot\}$ indicates vector representation. The existence of a scalar energy function forces $\mathbb{E}$ to be symmetric since the strains are symmetric, in other words $W = \frac{1}{2}\boldsymbol{\varepsilon} : \mathbb{E} : \boldsymbol{\varepsilon} = \frac{1}{2}(\boldsymbol{\varepsilon} : \mathbb{E} : \boldsymbol{\varepsilon})^T = \frac{1}{2}\boldsymbol{\varepsilon}^T : \mathbb{E}^T : \boldsymbol{\varepsilon}^T = \frac{1}{2}\boldsymbol{\varepsilon} : \mathbb{E}^T : \boldsymbol{\varepsilon}$, implying $\mathbb{E}^T = \mathbb{E}$. Consequently, $\mathbb{E}$ has only 21 free constants. The nonnegativity of W imposes the restriction that $\mathbb{E}$ remain positive definite. At this point, based on many factors that depend on the material microstructure, it can be shown that the components of $\mathbb{E}$ may be written in terms of anywhere between 21 and 2 independent parameters. We explore such concepts further via the ideas of elastic symmetry.

7.11.2 Material Symmetry

Transformation matrices are used in determining the symmetries. Consider a plane of symmetry, that is, the $x_2 - x_3$ plane (Figure 7.3). A plane of symmetry implies that the material has the same properties with respect to that plane. Therefore, we should be able to flip the axes with respect to that plane, and have no change in the constitutive law. By definition, a plane of elastic symmetry exists at a point where the elastic constants have the same value for a pair of coordinate systems. The axes are referred to as "equivalent elastic directions." Also, by definition, an axis of symmetry of order K

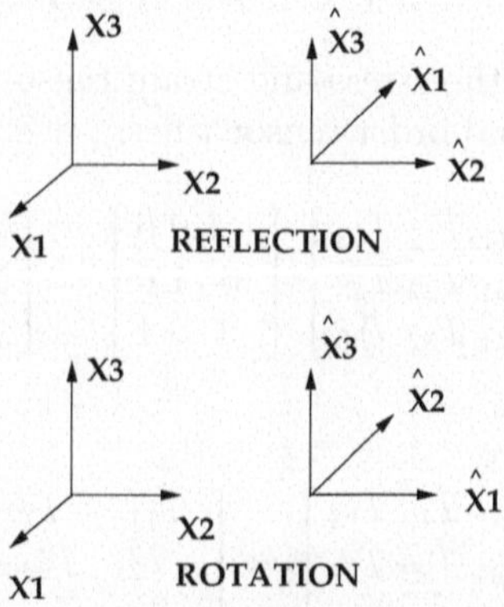

Fig. 7.3. Top: reflection with respect to the $x_2 - x_3$ plane. Bottom: rotation with respect to the x_3 axis.

exists at a point when there are sets of equivalent elastic directions which can be superposed by a rotation through an angle $2\pi/K$ about an axis. The way to determine elastic symmetry is as follows, $[\hat{\boldsymbol{T}}] = [\mathcal{T}]^{-1}[\boldsymbol{T}][\mathcal{T}]$ implying $\{\hat{\boldsymbol{T}}\} = [\mathcal{T}]\{\boldsymbol{T}\} = [\hat{\mathbb{E}}][\mathcal{T}]\{\boldsymbol{\varepsilon}\} = [\hat{\mathbb{E}}]\{\hat{\boldsymbol{\varepsilon}}\}$ which implies $\{\boldsymbol{T}\} = [\mathcal{T}^{-1}][\hat{\mathbb{E}}][\mathcal{T}]\{\boldsymbol{\varepsilon}\}$, where $[\boldsymbol{T}]$ is a rotational (or reflectional) transformation matrix. Imposing elastic symmetry means the components are invariant with respect to the transformation, therefore $[\mathbb{E}] = [\mathcal{T}^{-1}][\hat{\mathbb{E}}][\mathcal{T}]$. Therefore, all components that are not identical must be zero if the material has the assumed elastic symmetry. In this fashion one can "carve" away components from a general anisotropic material tensor. The central point of such symmetries is that in a new transformed state, $\mathbb{E}$ should not change.

Examples of elastic symmetry

To clarify, consider the following steps for one plane of symmetry starting with an originally triclinic material (21 free constants):

$$[\mathbb{E}] \stackrel{\text{def}}{=} \begin{bmatrix} E_{1111} & E_{1122} & E_{1133} & E_{1112} & E_{1123} & E_{1113} \\ E_{2211} & E_{2222} & E_{2233} & E_{2212} & E_{2223} & E_{2213} \\ E_{3311} & E_{3322} & E_{3333} & E_{3312} & E_{3323} & E_{3313} \\ E_{1211} & E_{1222} & E_{1233} & E_{1212} & E_{1223} & E_{1213} \\ E_{2311} & E_{2322} & E_{2333} & E_{2312} & E_{2323} & E_{2313} \\ E_{1311} & E_{1322} & E_{1333} & E_{1312} & E_{1323} & E_{1313} \end{bmatrix}. \quad (7.34)$$

- STEP 1: Reflect the x_1 axis

$$\boldsymbol{R}(x_1) \stackrel{\text{def}}{=} \begin{bmatrix} -1 & 0 & 0 \\ 0 & 1 & 0 \\ 0 & 0 & 1 \end{bmatrix} \begin{Bmatrix} x_1 \\ x_2 \\ x_3 \end{Bmatrix} = \begin{Bmatrix} \hat{x}_1 \\ \hat{x}_2 \\ \hat{x}_3 \end{Bmatrix} = \begin{Bmatrix} -x_1 \\ x_2 \\ x_3 \end{Bmatrix}. \tag{7.35}$$

- STEP 2: Transform the stress and strain tensors with the same transformation, but for second order tensor rules:

$$\begin{bmatrix} -1 & 0 & 0 \\ 0 & 1 & 0 \\ 0 & 0 & 1 \end{bmatrix} \begin{bmatrix} T_{11} & T_{12} & T_{13} \\ T_{21} & T_{22} & T_{23} \\ T_{31} & T_{32} & T_{33} \end{bmatrix} \begin{bmatrix} -1 & 0 & 0 \\ 0 & 1 & 0 \\ 0 & 0 & 1 \end{bmatrix} = \begin{bmatrix} \hat{T}_{11} & \hat{T}_{12} & \hat{T}_{13} \\ \hat{T}_{21} & \hat{T}_{22} & \hat{T}_{23} \\ \hat{T}_{31} & \hat{T}_{32} & \hat{T}_{33} \end{bmatrix}, \tag{7.36}$$

and thus

$$\begin{bmatrix} \hat{T}_{11} & \hat{T}_{12} & \hat{T}_{13} \\ \hat{T}_{21} & \hat{T}_{22} & \hat{T}_{23} \\ \hat{T}_{31} & \hat{T}_{32} & \hat{T}_{33} \end{bmatrix} = \begin{bmatrix} T_{11} & -T_{12} & -T_{13} \\ -T_{21} & T_{22} & T_{23} \\ -T_{31} & T_{32} & T_{33} \end{bmatrix}. \tag{7.37}$$

Also,

$$\begin{bmatrix} -1 & 0 & 0 \\ 0 & 1 & 0 \\ 0 & 0 & 1 \end{bmatrix} \begin{bmatrix} \varepsilon_{11} & \varepsilon_{12} & \varepsilon_{13} \\ \varepsilon_{21} & \varepsilon_{22} & \varepsilon_{23} \\ \varepsilon_{31} & \varepsilon_{32} & \varepsilon_{33} \end{bmatrix} \begin{bmatrix} -1 & 0 & 0 \\ 0 & 1 & 0 \\ 0 & 0 & 1 \end{bmatrix}, \tag{7.38}$$

and also

$$\begin{bmatrix} \hat{\varepsilon}_{11} & \hat{\varepsilon}_{12} & \hat{\varepsilon}_{13} \\ \hat{\varepsilon}_{21} & \hat{\varepsilon}_{22} & \hat{\varepsilon}_{23} \\ \hat{\varepsilon}_{31} & \hat{\varepsilon}_{32} & \hat{\varepsilon}_{33} \end{bmatrix} = \begin{bmatrix} \varepsilon_{11} & -\varepsilon_{12} & -\varepsilon_{13} \\ -\varepsilon_{21} & \varepsilon_{22} & \varepsilon_{23} \\ -\varepsilon_{31} & \varepsilon_{32} & \varepsilon_{33} \end{bmatrix}. \tag{7.39}$$

- STEP 3: Form the constitutive law in the primed frame:

$$\begin{Bmatrix} T_{11} \\ T_{22} \\ T_{33} \\ -T_{12} \\ T_{23} \\ -T_{31} \end{Bmatrix} = \begin{Bmatrix} \hat{T}_{11} \\ \hat{T}_{22} \\ \hat{T}_{33} \\ \hat{T}_{12} \\ \hat{T}_{23} \\ \hat{T}_{31} \end{Bmatrix}, \tag{7.40}$$

and

$$\begin{Bmatrix} \hat{T}_{11} \\ \hat{T}_{22} \\ \hat{T}_{33} \\ \hat{T}_{12} \\ \hat{T}_{23} \\ \hat{T}_{31} \end{Bmatrix} = \begin{bmatrix} \hat{E}_{1111} & \hat{E}_{1122} & \hat{E}_{1133} & \hat{E}_{1112} & \hat{E}_{1123} & \hat{E}_{1113} \\ \hat{E}_{2211} & \hat{E}_{2222} & \hat{E}_{2233} & \hat{E}_{2212} & \hat{E}_{2223} & \hat{E}_{2213} \\ \hat{E}_{3311} & \hat{E}_{3322} & \hat{E}_{3333} & \hat{E}_{3312} & \hat{E}_{3323} & \hat{E}_{3313} \\ \hat{E}_{1211} & \hat{E}_{1222} & \hat{E}_{1233} & \hat{E}_{1212} & \hat{E}_{1223} & \hat{E}_{1213} \\ \hat{E}_{2311} & \hat{E}_{2322} & \hat{E}_{2333} & \hat{E}_{2312} & \hat{E}_{2323} & \hat{E}_{2313} \\ \hat{E}_{1311} & \hat{E}_{1322} & \hat{E}_{1333} & \hat{E}_{1312} & \hat{E}_{1323} & \hat{E}_{1313} \end{bmatrix} \begin{Bmatrix} \hat{\varepsilon}_{11} \\ \hat{\varepsilon}_{22} \\ \hat{\varepsilon}_{33} \\ 2\hat{\varepsilon}_{12} \\ 2\hat{\varepsilon}_{23} \\ 2\hat{\varepsilon}_{31} \end{Bmatrix}. \tag{7.41}$$

- STEP 4: Put everything in terms of the original variables, which implies that the constitutive law must be the same as before (i.e., the components of $\mathbb{E}$ and $\hat{\mathbb{E}}$ must be the same) if the plane was a plane of symmetry, and thus the tensor relating $\boldsymbol{T}$ and $\boldsymbol{\varepsilon}$ is:

$$\begin{bmatrix} E_{1111} & E_{1122} & E_{1133} & -E_{1112} & E_{1123} & -E_{1113} \\ E_{2211} & E_{2222} & E_{2233} & -E_{2212} & E_{2223} & -E_{2213} \\ E_{3311} & E_{3322} & E_{3333} & -E_{3312} & E_{3323} & -E_{3313} \\ -E_{1211} & -E_{1222} & -E_{1233} & E_{1212} & -E_{1223} & E_{1213} \\ E_{2311} & E_{2322} & E_{2333} & -E_{2312} & E_{2323} & -E_{2313} \\ -E_{1311} & -E_{1322} & -E_{1333} & E_{1312} & -E_{1323} & E_{1313} \end{bmatrix}. \tag{7.42}$$

- STEP 5: All components that are not equal before and after the reflection of axes are zero:

$$\begin{Bmatrix} T_{11} \\ T_{22} \\ T_{33} \\ T_{12} \\ T_{23} \\ T_{31} \end{Bmatrix} = \begin{bmatrix} E_{1111} & E_{1122} & E_{1133} & 0 & E_{1123} & 0 \\ E_{2211} & E_{2222} & E_{2233} & 0 & E_{2223} & 0 \\ E_{3311} & E_{3322} & E_{3333} & 0 & E_{3323} & 0 \\ 0 & 0 & 0 & E_{1212} & 0 & E_{1213} \\ E_{2311} & E_{2322} & E_{2333} & 0 & E_{2323} & 0 \\ 0 & 0 & 0 & E_{1312} & 0 & E_{1313} \end{bmatrix} \begin{Bmatrix} \varepsilon_{11} \\ \varepsilon_{22} \\ \varepsilon_{33} \\ 2\varepsilon_{12} \\ 2\varepsilon_{23} \\ 2\varepsilon_{31} \end{Bmatrix}. \tag{7.43}$$

The end result is a monoclinic material, i.e, one plane of elastic symmetry ($x_2 - x_3$ plane)(13 free constants)

$$\mathbb{E} \stackrel{\text{def}}{=} \begin{bmatrix} E_{1111} & E_{1122} & E_{1133} & 0 & E_{1123} & 0 \\ E_{2211} & E_{2222} & E_{2233} & 0 & E_{2223} & 0 \\ E_{3311} & E_{3322} & E_{3333} & 0 & E_{3323} & 0 \\ 0 & 0 & 0 & E_{1212} & 0 & E_{1213} \\ E_{2311} & E_{2322} & E_{2333} & 0 & E_{2323} & 0 \\ 0 & 0 & 0 & E_{1312} & 0 & E_{1313} \end{bmatrix}, \tag{7.44}$$

or ($x_1 - x_3$ plane)(13 free constants)

$$\mathbb{E} \stackrel{\text{def}}{=} \begin{bmatrix} E_{1111} & E_{1122} & E_{1133} & 0 & 0 & E_{1113} \\ E_{2211} & E_{2222} & E_{2233} & 0 & 0 & E_{2213} \\ E_{3311} & E_{3322} & E_{3333} & 0 & 0 & E_{3313} \\ 0 & 0 & 0 & E_{1212} & E_{1223} & 0 \\ 0 & 0 & 0 & E_{1223} & E_{2323} & 0 \\ E_{1311} & E_{1322} & E_{1333} & 0 & 0 & E_{1313} \end{bmatrix}. \tag{7.45}$$

The basic procedure is the same, for reflectional symmetry, rotational symmetries, and so on. What follows is a catalog of commonly referred to materials of various symmetries.

- Two mutually perpendicular planes of symmetry reduce the material symmetry to nine free constants known as an *orthotropic* material. It also has a third mutually perpendicular plane of symmetry as a consequence, without changing the number of elastic constants. In other words, if one were to reflect a reference frame located at a material point (rotate by 180 degrees), the properties are the same. Accordingly, for an orthotropic material, there are two planes of symmetry (nine free constants)

$$\mathbb{E} \stackrel{\text{def}}{=} \begin{bmatrix} E_{1111} & E_{1122} & E_{1133} & 0 & 0 & 0 \\ E_{2211} & E_{2222} & E_{2233} & 0 & 0 & 0 \\ E_{3311} & E_{3322} & E_{3333} & 0 & 0 & 0 \\ 0 & 0 & 0 & E_{1212} & 0 & 0 \\ 0 & 0 & 0 & 0 & E_{2323} & 0 \\ 0 & 0 & 0 & 0 & 0 & E_{1313} \end{bmatrix}. \tag{7.46}$$

- In addition, if there is one plane in which the material properties are equal in all directions, there are only five free constants and the material is termed *transversely isotropic*. Accordingly, for transversely isotropic material: two planes of symmetry and one plane of directional independence (five free constants)

$$\mathbb{E} \stackrel{\text{def}}{=} \begin{bmatrix} E_{1111} & E_{1122} & E_{1133} & 0 & 0 & 0 \\ E_{2211} & E_{2222} & E_{2233} & 0 & 0 & 0 \\ E_{3311} & E_{3322} & E_{3333} & 0 & 0 & 0 \\ 0 & 0 & 0 & E_{1212} & 0 & 0 \\ 0 & 0 & 0 & 0 & E_{2323} & 0 \\ 0 & 0 & 0 & 0 & 0 & E_{1313} \end{bmatrix} \tag{7.47}$$

and $E_{1111} = E_{2222}$, $E_{1133} = E_{2233}$, $E_{1313} = E_{2323}$, $E_{1212} = \frac{1}{2}(E_{1111} - E_{1122})$.
- Cubic materials have two planes of symmetry and two planes of directional independence, and can be shown to have three free constants. Accordingly, for a cubic material: two planes of symmetry and two planes of directional independence (three free constants)

$$\mathbb{E} \stackrel{\text{def}}{=} \begin{bmatrix} E_{1111} & E_{1122} & E_{1133} & 0 & 0 & 0 \\ E_{2211} & E_{2222} & E_{2233} & 0 & 0 & 0 \\ E_{3311} & E_{3322} & E_{3333} & 0 & 0 & 0 \\ 0 & 0 & 0 & E_{1212} & 0 & 0 \\ 0 & 0 & 0 & 0 & E_{2323} & 0 \\ 0 & 0 & 0 & 0 & 0 & E_{1313} \end{bmatrix}, \tag{7.48}$$

and $E_{1111} = E_{2222} = E_{3333}$, $E_{1122} = E_{1133} = E_{2233}$, $E_{1313} = E_{2323} = E_{1212}$.
- Finally, if there are an infinite number of planes where the material properties are equal in all directions, there are two free constants, the Lamé parameters, and the material is of the familiar *isotropic* variety. An isotropic body has material properties that are the same in every direction at a point in the body, i.e., the properties are not a function of orientation at a point in a body. Accordingly, for isotropic materials: two planes of symmetry and an infinite number of planes of directional independence (two free constants),

$$\mathbb{E} \stackrel{\text{def}}{=} \begin{bmatrix} \kappa + \frac{4}{3}\overline{\mu} & \kappa - \frac{2}{3}\overline{\mu} & \kappa - \frac{2}{3}\overline{\mu} & 0 & 0 & 0 \\ \kappa - \frac{2}{3}\overline{\mu} & \kappa + \frac{4}{3}\overline{\mu} & \kappa - \frac{2}{3}\overline{\mu} & 0 & 0 & 0 \\ \kappa - \frac{2}{3}\overline{\mu} & \kappa - \frac{2}{3}\overline{\mu} & \kappa + \frac{4}{3}\overline{\mu} & 0 & 0 & 0 \\ 0 & 0 & 0 & \overline{\mu} & 0 & 0 \\ 0 & 0 & 0 & 0 & \overline{\mu} & 0 \\ 0 & 0 & 0 & 0 & 0 & \overline{\mu} \end{bmatrix}, \tag{7.49}$$

where κ is the bulk modulus and $\overline{\mu}$ is the shear modulus.

In this case, we have

$$\boxed{\mathbb{E} : \boldsymbol{\varepsilon} = 3\kappa \frac{tr\boldsymbol{\varepsilon}}{3}\mathbf{1} + 2\overline{\mu}\boldsymbol{\varepsilon}' \Rightarrow \boldsymbol{\varepsilon} : \mathbb{E} : \boldsymbol{\varepsilon} = 9\kappa(\frac{tr\boldsymbol{\varepsilon}}{3})^2 + 2\overline{\mu}\boldsymbol{\varepsilon}' : \boldsymbol{\varepsilon}',} \tag{7.50}$$

where $tr\boldsymbol{\varepsilon} = \varepsilon_{ii}$ and $\boldsymbol{\varepsilon}' = \boldsymbol{\varepsilon} - \frac{1}{3}(tr\boldsymbol{\varepsilon})\mathbf{1}$. The eigenvalues of an isotropic elasticity tensor are $(3\kappa, 2\overline{\mu}, 2\overline{\mu}, \overline{\mu}, \overline{\mu}, \overline{\mu})$. Therefore, we must have $\kappa > 0$ and $\overline{\mu} > 0$ to retain positive definiteness of $\mathbb{E}$.

Remark: Obviously, the elasticity tensor, regardless of the degree of anisotropy, must have positive eigenvalues. It is symmetric, and therefore, writing the similarity transform, the 6×6 collection of 6×1 mutually orthonormal eigenvectors of the elasticity as $\boldsymbol{Z}$ we have $[\boldsymbol{Z}^T][\mathbb{E}][\boldsymbol{Z}] = [\boldsymbol{\Lambda}]$, where $[\boldsymbol{\Lambda}]$ is diagonalized: $\{\boldsymbol{T}\} = [\boldsymbol{Z}][\boldsymbol{\Lambda}][\boldsymbol{Z}^T]\{\varepsilon\}$ which implies $[\boldsymbol{Z}^T]\{\boldsymbol{T}\} = \{\hat{\boldsymbol{T}}\} = [\boldsymbol{\Lambda}][\boldsymbol{Z}^T]\{\varepsilon\} = [\boldsymbol{\Lambda}]\{\hat{\varepsilon}\}$. Clearly the constitutive law in the new frame has the form:

$$\begin{Bmatrix} \hat{T}_{11} \\ \hat{T}_{22} \\ \hat{T}_{33} \\ \hat{T}_{12} \\ \hat{T}_{23} \\ \hat{T}_{13} \end{Bmatrix} = \begin{bmatrix} \Lambda_1 & 0 & 0 & 0 & 0 & 0 \\ 0 & \Lambda_2 & 0 & 0 & 0 & 0 \\ 0 & 0 & \Lambda_3 & 0 & 0 & 0 \\ 0 & 0 & 0 & \Lambda_4 & 0 & 0 \\ 0 & 0 & 0 & 0 & \Lambda_5 & 0 \\ 0 & 0 & 0 & 0 & 0 & \Lambda_6 \end{bmatrix} \begin{Bmatrix} \hat{\varepsilon}_{11} \\ \hat{\varepsilon}_{22} \\ \hat{\varepsilon}_{33} \\ 2\hat{\varepsilon}_{12} \\ 2\hat{\varepsilon}_{23} \\ 2\hat{\varepsilon}_{13} \end{Bmatrix}. \tag{7.51}$$

The strain energy is simply

$$W = \frac{1}{2}\left(\Lambda_1(\hat{\varepsilon}_{11})^2 + \Lambda_2(\hat{\varepsilon}_{22})^2 + \Lambda_3(\hat{\varepsilon}_{33})^2 + 4\Lambda_4(\hat{\varepsilon}_{12})^2 + 4\Lambda_5(\hat{\varepsilon}_{23})^2 + 4\Lambda_6(\hat{\varepsilon}_{13})^2\right). \tag{7.52}$$

One interpretation is that of a generalized decomposition of the energy into pieces associated with the principle directions of the the elasticity tensor in matrix form. Clearly, each eigenvalue, $\Lambda_i, i = 1, ..., 6$, must be positive.

7.11.3 Material Constant Interpretation

There are a variety of ways to write isotropic constitutive laws each time with a physically meaningful pair of material constants.

Splitting the strain

It is sometimes important to split infinitesimal strains into two physically meaningful parts $\boldsymbol{\varepsilon} = \frac{tr\mathcal{E}}{3}\mathbf{1} + (\boldsymbol{\varepsilon} - \frac{tr\mathcal{E}}{3}\mathbf{1})$. The Jacobian, J, of the deformation gradient $\boldsymbol{F}$ is $det(\mathbf{1} + \nabla_X \boldsymbol{u})$, and can be written as

$$\begin{vmatrix} \frac{\partial x_1}{\partial X_1} & \frac{\partial x_2}{\partial X_1} & \frac{\partial x_3}{\partial X_1} \\ \frac{\partial x_1}{\partial X_2} & \frac{\partial x_2}{\partial X_2} & \frac{\partial x_3}{\partial X_2} \\ \frac{\partial x_1}{\partial X_3} & \frac{\partial x_2}{\partial X_3} & \frac{\partial x_3}{\partial X_3} \end{vmatrix} dS^3 = \begin{vmatrix} 1+\frac{\partial u_1}{\partial X_1} & \frac{\partial u_2}{\partial X_1} & \frac{\partial u_3}{\partial X_1} \\ \frac{\partial u_1}{\partial X_2} & 1+\frac{\partial u_2}{\partial X_2} & \frac{\partial u_3}{\partial X_2} \\ \frac{\partial u_1}{\partial X_3} & \frac{\partial u_2}{\partial X_3} & 1+\frac{\partial u_3}{\partial X_3} \end{vmatrix} dS^3 \tag{7.53}$$

and expanded as $J = det\boldsymbol{F} = det(\mathbf{1} + \nabla_X \boldsymbol{u}) \approx 1 + tr\nabla_X \boldsymbol{u} + \mathcal{O}(\nabla_X \boldsymbol{u}) = 1 + tr\varepsilon + ...$, and therefore with infinitesimal strains $(1+tr\varepsilon)d\omega_0 = d\omega \Rightarrow tr\varepsilon = \frac{d\omega - d\omega_0}{d\omega_0}$. Hence, $tr\varepsilon$ is associated with the volumetric part of the deformation. Furthermore, since $tr(\boldsymbol{\varepsilon} - \frac{tr\mathcal{E}}{3}\mathbf{1}) = 0$, the so-called "strain deviator" can only affect the shape of a differential element. *In other words, it describes distortion in the material.*

Associated infinitesimal strain linearly elastic material laws

The stress $\boldsymbol{T}$ can be split into two parts (dilatational and a deviatoric): $\boldsymbol{T} = \frac{tr\boldsymbol{T}}{3}\mathbf{1} + (\boldsymbol{T} - \frac{tr\boldsymbol{T}}{3}\mathbf{1}) \stackrel{\text{def}}{=} p\mathbf{1} + \boldsymbol{T}'$, where we call the symbol p the hydrostatic pressure, and $\boldsymbol{T}'$ the stress deviator. This is one form of the so-called Hooke's Law. The resistance to change in the volume is measured by κ. We note that $(\frac{tr\boldsymbol{T}}{3}\mathbf{1})' = \mathbf{0}$, which indicates that this part of the stress produces no distortion. Another fundamental form of Hooke's law is $\boldsymbol{T} = \frac{E}{1+\nu}\left(\boldsymbol{\varepsilon} + \frac{\nu}{1-2\nu} tr\boldsymbol{\varepsilon}\mathbf{1}\right)$, which implies $\boldsymbol{\varepsilon} = \frac{1+\nu}{E}\boldsymbol{T} - \frac{\nu}{E} tr\boldsymbol{T}\mathbf{1}$. To interpret the constants, consider a uniaxial test where $T_{12} = T_{13} = T_{23} = 0 \Rightarrow \varepsilon_{12} = \varepsilon_{13} = \varepsilon_{23} = 0$, $T_{22} = T_{33} = 0$. Under these conditions we have $T_{11} = E\varepsilon_{11}$ and $\varepsilon_{22} = \varepsilon_{33} = -\nu\varepsilon_{11}$. Therefore, E, the so-called "Young's" modulus, is the ratio of the uniaxial stress to the corresponding strain component. The Poisson ratio, ν, is the ratio of the transverse strains to the uniaxial strain.

Another commonly used set of stress-strain forms are the Lamé relations, $\boldsymbol{T} = \lambda tr\boldsymbol{\varepsilon}\mathbf{1} + 2\overline{\mu}\boldsymbol{\varepsilon}$ or $\boldsymbol{\varepsilon} = -\frac{\lambda}{2\overline{\mu}(3\lambda+2\overline{\mu})} tr\boldsymbol{T}\mathbf{1} + \frac{\boldsymbol{T}}{2\overline{\mu}}$. To interpret the constants, consider a pressure test where $T_{12} = T_{13} = T_{23} = 0$, and where $T_{11} = T_{22} = T_{33}$. Under these conditions we have $\kappa = \lambda + \frac{2}{3}\overline{\mu} = \frac{E}{3(1-2\nu)}$, $\overline{\mu} = \frac{E}{2(1+\nu)}$ and $\frac{\kappa}{\overline{\mu}} = \frac{2(1+\nu)}{3(1-2\nu)}$. We observe that $\frac{\kappa}{\overline{\mu}} \to \infty$ which implies that $\nu \to \frac{1}{2}$, and $\frac{\kappa}{\overline{\mu}} \to 0$ implies $\Rightarrow \nu \to -1$. Therefore, from the fact that both κ and $\overline{\mu}$ must be positive and finite, this implies $-1 < \nu < 0.5$ and $0 < E < \infty$. For example, some polymeric foams exhibit $\nu < 0$, steels $\nu \approx 0.3$, and some forms of rubber have $\nu \to 0.5$. However, *no restrictions arise on λ, i.e. it could be positive or negative.*

8

Basic Time-Stepping Schemes

Generally, methods for the time integration of differential equations falls within two broad categories: (1) implicit and (2) explicit. In order to clearly delineate between the two approaches, we first study a generic equation of the form ($\boldsymbol{U}$=unknown variable)

$$\mathcal{M}\dot{\boldsymbol{U}}(t) = \boldsymbol{F}(\boldsymbol{U}, t) \Rightarrow \dot{\boldsymbol{U}}(t) = \boldsymbol{\mathcal{G}}(\boldsymbol{U}, t). \tag{8.1}$$

If we discretize the differential equation in the following manner, that is,

$$\dot{\boldsymbol{U}}(t) \approx \frac{\boldsymbol{U}(t+\Delta t) - \boldsymbol{U}(t)}{\Delta t} = \boldsymbol{\mathcal{G}}(\boldsymbol{U}(t), t) \Rightarrow \boldsymbol{U}(t+\Delta t) = \boldsymbol{U}(t) + \Delta t \boldsymbol{\mathcal{G}}(\boldsymbol{U}(t), t), \tag{8.2}$$

we obtain an *explicit* expression for $\boldsymbol{U}(t+\Delta t)$. This is often referred to as a *Forward Euler* scheme. If we discretize the differential equation as

$$\dot{\boldsymbol{U}}(t+\Delta t) = \frac{\boldsymbol{U}(t+\Delta t) - \boldsymbol{U}(t)}{\Delta t} = \boldsymbol{\mathcal{G}}(\boldsymbol{U}(t+\Delta t), t+\Delta t), \tag{8.3}$$

we obtain

$$\boldsymbol{U}(t+\Delta t) = \boldsymbol{U}(t) + \Delta t \boldsymbol{\mathcal{G}}(\boldsymbol{U}(t+\Delta t), t+\Delta t), \tag{8.4}$$

which yields an *implicit* expression, which can be nonlinear in $\boldsymbol{U}(t+\Delta t)$, depending on $\boldsymbol{\mathcal{G}}$. This is often referred to as a *Backward Euler* scheme. These two techniques illustrate the most basic time-stepping schemes used in the scientific community, which form the foundation for the majority of more sophisticated methods. Two main observations can be made:

- The implicit method usually requires that one solves a (potentially nonlinear) equation for $\boldsymbol{U}(t+\Delta t)$.
- The explicit method has a major drawback, namely, that the step size Δt may have to be very small to achieve acceptable numerical results. Therefore, an explicit simulation will usually require many more time steps than an implicit simulation for the same quality of results.

T.I. Zohdi: Electromagnetic Properties of Multiphase Dielectrics, LNACM 64, pp. 89–101.
springerlink.com

- Generally speaking, a key difference between explicit and implicit schemes are their stability properties. By *stability*, we mean that errors made at one stage of the calculations do not cause increasingly larger errors as the computations are continued. Most time-stepping schemes may be collapsed to the following form:

$$\boldsymbol{U}(t+\Delta t) = \boldsymbol{A}\cdot\boldsymbol{U}(t) = \underbrace{(\boldsymbol{A}\cdot\boldsymbol{A}\cdot\boldsymbol{A}\cdot\boldsymbol{A}...)}_{\stackrel{\text{def}}{=}\boldsymbol{A}^N}\cdot\boldsymbol{U}(t_o). \tag{8.5}$$

Defining the initial error $\boldsymbol{\varpi}(t_o) = \boldsymbol{U}(t_o) - \boldsymbol{U}^*(t_o)$, where $\boldsymbol{U}^*$ is the true solution and the error at time t,

$$\boldsymbol{\varpi}(t) = \boldsymbol{U}(t) - \boldsymbol{U}^*(t) = \boldsymbol{A}^N\cdot\boldsymbol{U}(t_o) - \boldsymbol{A}^N\cdot\boldsymbol{U}^*(t_o) = \boldsymbol{A}^N\cdot\boldsymbol{\varpi}(t_o), \tag{8.6}$$

if the magnitude of the maximum eigenvalue of $\boldsymbol{A}$ is greater than unity, and thus the error will increase. For illustration purposes, consider applying each method to the linear scalar differential equation

$$\dot{\boldsymbol{U}} = -c\boldsymbol{U}, \tag{8.7}$$

where $\boldsymbol{U}(0) = \boldsymbol{U}_o$ and c is a positive constant. The exact solution is $\boldsymbol{U}(t) = \boldsymbol{U}_o e^{-ct}$. For the explicit method,

$$\dot{\boldsymbol{U}} \approx \frac{\boldsymbol{U}(t+\Delta t) - \boldsymbol{U}(t)}{\Delta t} = -c\boldsymbol{U}(t), \tag{8.8}$$

which leads to the following time-stepping scheme, that is,

$$\boldsymbol{U}(L\Delta t) = \boldsymbol{U}_o(1 - c\Delta t)^L, \tag{8.9}$$

where L indicates the time step counter, namely, $t = L\Delta t$ for uniform time steps (as in this example) and $\boldsymbol{U}^L \stackrel{\text{def}}{=} \boldsymbol{U}(t)$, and so on. It is stable if $|1 - c\Delta t| < 1$. For the implicit method,

$$\dot{\boldsymbol{U}} \approx \frac{\boldsymbol{U}(t+\Delta t) - \boldsymbol{U}(t)}{\Delta t} = -c\boldsymbol{U}(t+\Delta t), \tag{8.10}$$

which leads to the following time-stepping scheme

$$\boldsymbol{U}(L\Delta t) = \frac{\boldsymbol{U}_o}{(1 + c\Delta t)^L}. \tag{8.11}$$

Since $\frac{1}{1+c\Delta t} < 1$, it is always stable. Note that the approximation in Equation 8.9 oscillates in an artificial, nonphysical manner when

$$\Delta t > \frac{2}{c}. \tag{8.12}$$

If $c >> 1$, then Equation 8.7 is a so-called "stiff" equation, and Δt may have to be very small for the explicit method to be stable, while, for this example, a larger value of Δt can be used with the implicit method. This motivates the use of implicit methods with adaptive time-stepping which will be used throughout the remaining analysis. *However, there are some advantages to combining explicit and implicit methods together, which we will discuss next.*

8.1 Hybrid Methods

Consider a model problem:

$$\mathcal{M}\dot{\boldsymbol{U}} = \boldsymbol{F}(\boldsymbol{U}). \tag{8.13}$$

Expanding $\boldsymbol{U}$ in a Taylor series about $t + \phi\Delta t$, we obtain[1]

$$\boldsymbol{U}(t+\Delta t) = \boldsymbol{U}(t+\phi\Delta t)+\dot{\boldsymbol{U}}|_{t+\phi\Delta t}(1-\phi)\Delta t+\frac{1}{2}\ddot{\boldsymbol{U}}|_{t+\phi\Delta t}(1-\phi)^2(\Delta t)^2+\mathcal{O}((\Delta t)^3) \tag{8.14}$$

and

$$\boldsymbol{U}(t) = \boldsymbol{U}(t + \phi\Delta t) - \dot{\boldsymbol{U}}|_{t+\phi\Delta t}\phi\Delta t + \frac{1}{2}\ddot{\boldsymbol{U}}|_{t+\phi\Delta t}\phi^2(\Delta t)^2 + \mathcal{O}((\Delta t)^3). \tag{8.15}$$

Subtracting the two expressions yields

$$\dot{\boldsymbol{U}}|_{t+\phi\Delta t} = \frac{\boldsymbol{U}(t + \Delta t) - \boldsymbol{U}(t)}{\Delta t} + \hat{\mathcal{O}}(\Delta t), \tag{8.16}$$

where $\hat{\mathcal{O}}(\Delta t) = \mathcal{O}((\Delta t)^2)$ when $\phi = \frac{1}{2}$, and $\hat{\mathcal{O}}(\Delta t) = \mathcal{O}(\Delta t)$ when $\phi \neq \frac{1}{2}$. Inserting this into the governing equations yields

$$\boldsymbol{U}(t + \Delta t) = \boldsymbol{U}(t) + \frac{\Delta t}{\mathcal{M}}\boldsymbol{F}(t + \phi\Delta t) + \hat{\mathcal{O}}((\Delta t)^2). \tag{8.17}$$

Note that adding a weighted sum of Equations 8.14 and 8.15 yields

$$\boldsymbol{U}(t + \phi\Delta t) = \phi\boldsymbol{U}(t + \Delta t) + (1 - \phi)\boldsymbol{U}(t) + \mathcal{O}((\Delta t)^2), \tag{8.18}$$

which will be useful shortly. The term $\boldsymbol{F}(t+\phi\Delta t)$ can be handled in two main ways:

- $\boldsymbol{F}(\boldsymbol{U}(t + \phi\Delta t)) \approx \boldsymbol{F}(\phi\boldsymbol{U}(t + \Delta t) + (1 - \phi)\boldsymbol{U}(t))$ or
- $\boldsymbol{F}(\boldsymbol{U}(t + \phi\Delta t)) \approx \phi\boldsymbol{F}(\boldsymbol{U}(t + \Delta t)) + (1 - \phi)\boldsymbol{F}(\boldsymbol{U}(t))$.

[1] A general rule to remember when constructing more advanced time-stepping formulae is the "expand around the time step you wish to evaluate at."

For brevity, we will take the latter approximation. In summary,

$$\begin{aligned}\boldsymbol{U}(t+\Delta t) &= \boldsymbol{U}(t) + \tfrac{\Delta t}{\mathcal{M}}\boldsymbol{F}(t+\phi\Delta t) + \hat{\mathcal{O}}((\Delta t)^2)\\ &= \boldsymbol{U}(t) + \tfrac{\Delta t}{\mathcal{M}}\left(\phi\boldsymbol{F}(\boldsymbol{U}(t+\Delta t)) + (1-\phi)\boldsymbol{F}(\boldsymbol{U}(t))\right) + \hat{\mathcal{O}}((\Delta t)^2).\end{aligned} \tag{8.19}$$

We note that

- When $\phi = 1$, then this is the (implicit) Backward Euler scheme, which is very stable (very dissipative) and $\mathcal{O}((\Delta t)^2)$ locally in time,
- When $\phi = 0$, then this is the (explicit) Forward Euler scheme, which is conditionally stable and $\mathcal{O}((\Delta t)^2)$ locally in time,
- When $\phi = 0.5$, then this is the (implicit) "Midpoint" scheme, which is marginally stable and $\hat{\mathcal{O}}((\Delta t)^2) = \mathcal{O}((\Delta t)^3)$ locally in time.

Remark 1: The term *convergence* pertains to whether the method approximates the solution

$$Lim_{\Delta t \to 0} max_{N=1,2,...} ||\boldsymbol{U}(t_N) - \boldsymbol{U}^*(t_N)|| = 0. \tag{8.20}$$

If a method is defined by

$$\boldsymbol{U}(t_N) = \boldsymbol{\Lambda}(t_N, \boldsymbol{U}(t_{N-1}..., \Delta t)), \tag{8.21}$$

and we define the local error as

$$\varpi(t_N) \stackrel{\text{def}}{=} \boldsymbol{\Lambda}(t_N) - \boldsymbol{U}^*(t_N), \tag{8.22}$$

the method is called *consistent* if[2]

$$Lim_{\Delta t \to 0} = \frac{\varpi(t_N)}{\Delta t} = 0. \tag{8.23}$$

The method is of order p if $\boldsymbol{U}(t_N) = \mathcal{O}((\Delta t)^{p+1})$ as $\Delta t \to 0$. The global error for N steps where $\Delta t = T/N$, $\boldsymbol{U}(t) - \boldsymbol{U}^*(t) = \mathcal{O}((\Delta t)^{p+1}))$, and multiplying by N steps yields $\mathcal{O}((\Delta t^{p+1}))N = \mathcal{O}((\Delta t)^p)$.

Remark 2: The numerical analysis of differential equations typically entails three issues:

- *Convergence*: whether the method approximates a solution and in the limit, as $\Delta t \to 0$, captures it exactly.[3] A numerical method is said to be convergent if the numerical solution approaches the exact solution as the step size $\Delta t \to 0$

---

[2] Generally, if a method is consistent and stable, it is convergent.

[3] A usual requirement is that when the equation is put in the form

$$\dot{\boldsymbol{U}} = \boldsymbol{\mathcal{G}}(\boldsymbol{U}), \tag{8.24}$$

that the solution satisfies a Lipschitz condition $||\boldsymbol{\mathcal{G}}(\boldsymbol{U}) - \boldsymbol{\mathcal{G}}(\boldsymbol{U}^*)|| \le C||\boldsymbol{U} - \boldsymbol{U}^*||$, where $0 \le C < \infty$. See any good book on numerical analysis for details.

$$\lim_{\Delta t \to 0} max_{t \in T} ||\boldsymbol{U} - \boldsymbol{U}^*|| = 0. \tag{8.25}$$

- *Order*: how well the method approximates the solution for finite Δt.
- *Stability*: whether errors are damped out.

8.2 Coupled Problems and Staggering Schemes

A convenient way of solving coupled systems of differential equations is by using recursive (iterative) staggering schemes. Generally, staggering schemes proceed by solving each differential equation individually, allowing only one variable to be active per equation, and "freezing" all others. This (momentarily) uncouples the system. After the solution of each differential equation, the active variable is updated, and the next differential equation is addressed in a similar manner. For example, consider an abstract setting for two equations:

$$\mathcal{A}_1(\boldsymbol{U}_1, \boldsymbol{U}_2) = \mathcal{B}_1(\boldsymbol{U}_1, \boldsymbol{U}_2) \tag{8.26}$$

and

$$\mathcal{A}_2(\boldsymbol{U}_1, \boldsymbol{U}_2) = \mathcal{B}_2(\boldsymbol{U}_1, \boldsymbol{U}_2). \tag{8.27}$$

Now, consider the following iterative scheme, where $K = 1, 2, ...N$, where first $\boldsymbol{U}_1$ is solved for, freezing $\boldsymbol{U}_2$

$$\mathcal{A}_1(\boldsymbol{U}_1^K, \boldsymbol{U}_2^{K-1}) = \mathcal{B}_1(\boldsymbol{U}_1^{K-1}, \boldsymbol{U}_2^{K-1}). \tag{8.28}$$

Then, $\boldsymbol{U}_1$ is updated and $\boldsymbol{U}_2$ is solved for, freezing $\boldsymbol{U}_1$

$$\mathcal{A}_2(\boldsymbol{U}_1^K, \boldsymbol{U}_2^K) = \mathcal{B}_2(\boldsymbol{U}_1^K, \boldsymbol{U}_2^{K-1}). \tag{8.29}$$

For the class of coupled systems considered in this work, the coupled operator's spectral radius (largest eigenvalue) controls the convergence (within a time step) of this type of method, and is directly dependent on the time step discretization size, Δt. We consider a simple example which illustrates the essential concepts. Accordingly, consider the coupling of two first order equations

$$\boxed{\begin{aligned} a\dot{U}_1 + cU_2 &= F_1, \\ b\dot{U}_2 + dU_1 &= F_2, \end{aligned}}. \tag{8.30}$$

When discretized in time, for example, with a ϕ-method $(0 \leq \phi \leq 1)$ scheme, we obtain

$$\boxed{\begin{array}{l} U_1^{L+1} = U_1^L + \frac{\Delta t}{a}\left(\phi(F_1^{L+1} - cU_2^{L+1}) + (1-\phi)(F_1^L - U_2^L)\right) \\ U_2^{L+1} = U_2^L + \frac{\Delta t}{b}\left(\phi(F_2^{L+1} - dU_1^{L+1}) + (1-\phi)(F_2^L - U_1^L)\right) \end{array}}, \tag{8.31}$$

leading to the following coupled system:

$$\begin{bmatrix} 1 & \frac{\phi \Delta t c}{a} \\ \frac{\phi \Delta t d}{b} & 1 \end{bmatrix} \begin{Bmatrix} U_1^{L+1} \\ U_2^{L+1} \end{Bmatrix} = \begin{Bmatrix} \underbrace{U_1^L + \frac{\Delta t}{a}\left(\phi F_1^{L+1} + (1-\phi)(F_1^L - cU_2^L)\right)}_{\stackrel{\text{def}}{=} \gamma_1} \\ \underbrace{U_2^L + \frac{\Delta t}{b}\left(\phi F_2^{L+1} + (1-\phi)(F_2^L - dU_1^L)\right)}_{\stackrel{\text{def}}{=} \gamma_2} \end{Bmatrix}. \tag{8.32}$$

Considering a recursive staggering scheme of Jacobi-type, where the updates are made only after one complete iteration, considered here only for algebraic simplicity, one has the following split[4]

$$\begin{bmatrix} 0 & \frac{\phi \Delta t c}{a} \\ \frac{\phi \Delta t d}{b} & 0 \end{bmatrix} \begin{Bmatrix} U_1^{L+1} \\ U_2^{L+1} \end{Bmatrix} + \begin{bmatrix} 1 & 0 \\ 0 & 1 \end{bmatrix} \begin{Bmatrix} U_1^{L+1} \\ U_2^{L+1} \end{Bmatrix} = \begin{Bmatrix} \gamma_1 \\ \gamma_2 \end{Bmatrix} \tag{8.33}$$

and rearranging for iterative solution

$$\underbrace{-\begin{bmatrix} 0 & \frac{\phi \Delta t c}{a} \\ \frac{\phi \Delta t d}{b} & 0 \end{bmatrix} \begin{Bmatrix} U_1^{L+1} \\ U_2^{L+1} \end{Bmatrix}}_{\mathcal{G}(\boldsymbol{U}^{L+1,K})} + \underbrace{\begin{Bmatrix} \gamma_1 \\ \gamma_2 \end{Bmatrix}}_{\mathcal{R}} = \underbrace{\begin{Bmatrix} U_1^{L+1} \\ U_2^{L+1} \end{Bmatrix}}_{\boldsymbol{U}^{L+1}}. \tag{8.34}$$

The eigenvalues of $\mathcal{G}$ are

$$\lambda_{1,2} = \pm\sqrt{\frac{(\phi \Delta t)^2 cd}{ab}}. \tag{8.35}$$

One sees that the convergence of the staggering scheme is directly related (linearly in this case) to the size of the time step by setting $|\lambda_{1,2}| < 1$, and we obtain

$$\Delta t < \frac{1}{\phi}\sqrt{\frac{ab}{cd}}. \tag{8.36}$$

[4] A Gauss–Seidel-type approach would involve using the most current iterate. Typically, under very general conditions, if the Jacobi method converges, the Gauss–Seidel method converges at a faster rate, while if the Jacobi method diverges, the Gauss–Seidel method diverges at a faster rate. For example, see Ames [3] for details. The Jacobi method is easier to address theoretically, and thus it is used for proof of convergence, and the Gauss–Seidel method at the implementation level.

Clearly, the number of iterations increase with ϕ; the least number of iterations (zero) being when an explicit Euler method is used, and the largest number of iterations being when an implicit Euler method ($\phi = 1$) is used. As pointed out in Zohdi [129–134], the time step induced restriction for convergence matches the radius of analyticity of a Taylor series expansion of the solution around time t.

Remark: The term $\sqrt{\frac{cd}{ab}}$ appears in the exact solution which contains exponentials of the form $Ae^{\sqrt{\frac{cd}{ab}}t}$ that dictate the rate of growth of the solution. Thus, the time step dictated by Equation 8.36 must be small enough to "compensate" for this growth.

8.3 Temporally-Adaptive Iterative Methods

Implicit time-stepping methods, with time step size adaptivity, built on approaches found in Zohdi [129–134], will be used throughout the following analysis. In order to introduce basic concepts, we consider a first order differential equation (application to Maxwell's equations will follow)

$$\mathcal{M}\dot{\boldsymbol{U}} = \boldsymbol{F}(\boldsymbol{U}), \tag{8.37}$$

which, after being discretized using a trapezoidal "ϕ-method" ($0 \leq \phi \leq 1$)

$$\boldsymbol{U}^{L+1} = \boldsymbol{U}^{L} + \frac{\Delta t}{\mathcal{M}}\left(\phi \boldsymbol{F}(\boldsymbol{U}^{L+1}) + (1-\phi)\boldsymbol{F}(\boldsymbol{U}^{L})\right), \tag{8.38}$$

yields the following abstract form

$$\mathcal{A}(\boldsymbol{U}^{L+1}) = \mathcal{B}. \tag{8.39}$$

It is convenient to write

$$\mathcal{A}(\boldsymbol{U}^{L+1}) - \mathcal{B} = \boldsymbol{\mathcal{G}}(\boldsymbol{U}^{L+1}) - \boldsymbol{U}^{L+1} + \boldsymbol{\mathcal{R}} = \boldsymbol{0}, \tag{8.40}$$

where $\boldsymbol{\mathcal{R}}$ is a remainder term that does not depend on the solution, i.e., $\boldsymbol{\mathcal{R}} \neq \boldsymbol{\mathcal{R}}(\boldsymbol{U}^{L+1})$. A straightforward iterative scheme can be written as

$$\boldsymbol{U}^{L+1,K} = \boldsymbol{\mathcal{G}}(\boldsymbol{U}^{L+1,K-1}) + \boldsymbol{\mathcal{R}}, \tag{8.41}$$

where $K = 1, 2, 3, ...$ is the index of iteration within time step $L + 1$. The convergence of such a scheme is dependent on the behavior of $\boldsymbol{\mathcal{G}}$. Namely, a sufficient condition for convergence is that $\boldsymbol{\mathcal{G}}$ is a contraction mapping for all $\boldsymbol{U}^{L+1,K}$, $K = 1, 2, 3...$ In order to investigate this further, we define the iteration error as

$$\boldsymbol{\varpi}^{L+1,K} \stackrel{\text{def}}{=} \boldsymbol{U}^{L+1,K} - \boldsymbol{U}^{L+1}. \tag{8.42}$$

A necessary restriction for convergence is iterative self consistency, i.e., the "exact" (discretized) solution must be represented by the scheme

$$\mathcal{G}(\boldsymbol{U}^{L+1}) + \boldsymbol{\mathcal{R}} = \boldsymbol{U}^{L+1}. \tag{8.43}$$

Enforcing this restriction, a sufficient condition for convergence is the existence of a contraction mapping

$$\begin{aligned} \varpi^{L+1,K} = ||\boldsymbol{U}^{L+1,K} - \boldsymbol{U}^{L+1}|| &= ||\mathcal{G}(\boldsymbol{U}^{L+1,K-1}) - \mathcal{G}(\boldsymbol{U}^{L+1})|| \\ &\leq \eta^{L+1,K}||\boldsymbol{U}^{L+1,K-1} - \boldsymbol{U}^{L+1}||, \end{aligned} \tag{8.44}$$

where, if $0 \leq \eta^{L+1,K} < 1$ for each iteration K, then $\varpi^{L+1,K} \to 0$ for any arbitrary starting value $\boldsymbol{U}^{L+1,K=0}$, as $K \to \infty$. This type of contraction condition is sufficient, but not necessary, for convergence. Inserting these approximations into $\mathcal{M}\dot{\boldsymbol{U}} = \boldsymbol{F}(\boldsymbol{U})$ leads to

$$\boldsymbol{U}^{L+1,K} \approx \underbrace{\frac{\Delta t}{\mathcal{M}}\left(\phi \boldsymbol{F}(\boldsymbol{U}^{L+1,K-1})\right)}_{\mathcal{G}(\boldsymbol{U}^{L+1,K-1})} + \underbrace{\frac{\Delta t}{\mathcal{M}}(1-\phi)\boldsymbol{F}(\boldsymbol{U}^{L}) + \boldsymbol{U}^{L}}_{\boldsymbol{\mathcal{R}}}, \tag{8.45}$$

whose contraction constant is scaled by $\eta \propto \frac{\phi \Delta t}{\mathcal{M}}$. Therefore, we see that the contraction constant η is (1) directly dependent on the strength of the interaction forces (implicitly represented via the magnitude of the operator, $||\boldsymbol{F}||$), (2) inversely proportional to $\mathcal{M}$ and (3) directly proportional to $\phi\Delta t$ (at time=t). Therefore, if convergence is slow within a time step, the time step size, which is adjustable, can be reduced by an appropriate amount to increase the rate of convergence. Decreasing the time step size improves the convergence, however, we want to simultaneously maximize the time step sizes to decrease overall computing time, while still meeting an error tolerance on the numerical solution's accuracy. In order to achieve this goal, we follow an approach found in Zohdi [129-134] originally developed for continuum thermochemical multifield problems in which one first approximates

$$\eta^{L+1,K} \approx S(\Delta t)^p, \tag{8.46}$$

(S is a constant) and secondly, one assumes the error within an iteration to behave according to

$$(S(\Delta t)^p)^K \varpi^{L+1,0} = \varpi^{L+1,K}, \tag{8.47}$$

$K = 1, 2, ...$, where $\varpi^{L+1,0}$ is the initial norm of the iterative error and S is intrinsic to the system.[5] Our goal is to meet an error tolerance in exactly a preset number of iterations. To this end, one writes

[5] For the class of problems under consideration, due to the quadratic dependency on Δt, $p \approx 1$.

$$(S(\Delta t_{\rm tol})^p)^{K_d} \varpi^{L+1,0} = C_{tol}, \tag{8.48}$$

where C_{tol} is a (coupling) tolerance and where K_d is the number of desired iterations.[6] If the error tolerance is not met in the desired number of iterations, the contraction constant $\eta^{L+1,K}$ is too large. Accordingly, one can solve for a new smaller step size, under the assumption that S is constant,

$$\Delta t_{\rm tol} = \Delta t \left(\frac{\left(\frac{C_{tol}}{\varpi^{L+1,0}}\right)^{\frac{1}{pK_d}}}{\left(\frac{\varpi^{L+1,K}}{\varpi^{L+1,0}}\right)^{\frac{1}{pK}}} \right). \tag{8.49}$$

The assumption that S is constant is not critical since the time steps are to be recursively refined and unrefined throughout the simulation. Clearly, the expression in Equation 8.49 can also be used for time step enlargement, if convergence is met in less than K_d iterations.[7]

8.4 Temporally Second Order Equations

Later, we will also need to discretize temporally second order differential equations (governing the balance of linear momentum). In order to clearly explain the time-stepping scheme, we first start with the dynamics of a particle. The equation of motion is given by

$$m\dot{\boldsymbol{v}} = \boldsymbol{\Psi}, \tag{8.50}$$

where $\boldsymbol{\Psi}$ is the total force provided from interactions with the external environment. Expanding the velocity in a Taylor series about $t + \phi\Delta t$, we obtain

$$\boldsymbol{v}(t+\Delta t) = \boldsymbol{v}(t+\phi\Delta t) + \frac{d\boldsymbol{v}}{dt}|_{t+\phi\Delta t}(1-\phi)\Delta t + \frac{1}{2}\frac{d^2\boldsymbol{v}}{dt^2}|_{t+\phi\Delta t}(1-\phi)^2(\Delta t)^2 + \mathcal{O}((\Delta t)^3) \tag{8.51}$$

and

$$\boldsymbol{v}(t) = \boldsymbol{v}(t+\phi\Delta t) - \frac{d\boldsymbol{v}}{dt}|_{t+\phi\Delta t}\phi\Delta t + \frac{1}{2}\frac{d^2\boldsymbol{v}}{dt^2}|_{t+\phi\Delta t}\phi^2(\Delta t)^2 + \mathcal{O}((\Delta t)^3). \tag{8.52}$$

Subtracting the two expressions yields

$$\frac{d\boldsymbol{v}}{dt}|_{t+\phi\Delta t} = \frac{\boldsymbol{v}(t+\Delta t) - \boldsymbol{v}(t)}{\Delta t} + \hat{\mathcal{O}}(\Delta t), \tag{8.53}$$

[6] Typically, K_d is chosen to be between five to ten iterations, although this is problem and analyst dependent.

[7] At the implementation level, since the exact solution is unknown, the following relative error term is used, $\varpi^{L+1,K} \stackrel{\rm def}{=} \boldsymbol{U}^{L+1,K} - \boldsymbol{U}^{L+1,K-1}$.

where $\hat{\mathcal{O}}(\Delta t) = \mathcal{O}((\Delta t)^2)$, when $\phi = \frac{1}{2}$. Thus, inserting this into the equations of equilibrium yields

$$\boldsymbol{v}(t + \Delta t) = \boldsymbol{v}(t) + \frac{\Delta t}{m}\boldsymbol{\Psi}(t + \phi\Delta t) + \hat{\mathcal{O}}((\Delta t)^2). \quad (8.54)$$

Note that adding a weighted sum of Equations 8.51 and 8.52 yields

$$\boldsymbol{v}(t + \phi\Delta t) = \phi\boldsymbol{v}(t + \Delta t) + (1 - \phi)\boldsymbol{v}(t) + \mathcal{O}((\Delta t)^2), \quad (8.55)$$

which will be useful shortly. Now, expanding the position of the center of mass in a Taylor series about $t + \phi\Delta t$, we obtain

$$\boldsymbol{u}(t+\Delta t) = \boldsymbol{u}(t+\phi\Delta t) + \frac{d\boldsymbol{u}}{dt}|_{t+\phi\Delta t}(1-\phi)\Delta t + \frac{1}{2}\frac{d^2\boldsymbol{u}}{dt^2}|_{t+\phi\Delta t}(1-\phi)^2(\Delta t)^2 + \mathcal{O}((\Delta t)^3) \quad (8.56)$$

and

$$\boldsymbol{u}(t) = \boldsymbol{u}(t+\phi\Delta t) - \frac{d\boldsymbol{u}}{dt}|_{t+\phi\Delta t}\phi\Delta t + \frac{1}{2}\frac{d^2\boldsymbol{u}}{dt^2}|_{t+\phi\Delta t}\phi^2(\Delta t)^2 + \mathcal{O}((\Delta t)^3). \quad (8.57)$$

Subtracting the two expressions yields

$$\frac{\boldsymbol{u}(t + \Delta t) - \boldsymbol{u}(t)}{\Delta t} = \boldsymbol{v}(t + \phi\Delta t) + \hat{\mathcal{O}}(\Delta t). \quad (8.58)$$

Inserting Equation 8.55 yields

$$\boldsymbol{u}(t + \Delta t) = \boldsymbol{u}(t) + (\phi\boldsymbol{v}(t + \Delta t) + (1 - \phi)\boldsymbol{v}(t))\,\Delta t + \hat{\mathcal{O}}((\Delta t)^2), \quad (8.59)$$

and thus using Equation 8.54 yields

$$\boldsymbol{u}(t + \Delta t) = \boldsymbol{u}(t) + \boldsymbol{v}(t)\Delta t + \frac{\phi(\Delta t)^2}{m}\boldsymbol{\Psi}(t + \phi\Delta t) + \hat{\mathcal{O}}((\Delta t)^2). \quad (8.60)$$

The term $\boldsymbol{\Psi}(t + \phi\Delta t)$ can be handled in two main ways:

- $\boldsymbol{\Psi}(t + \phi\Delta t) \approx \boldsymbol{\Psi}(\phi\boldsymbol{u}(t + \Delta t) + (1 - \phi)\boldsymbol{u}(t))$ or
- $\boldsymbol{\Psi}(t + \phi\Delta t) \approx \phi\boldsymbol{\Psi}(\boldsymbol{u}(t + \Delta t)) + (1 - \phi)\boldsymbol{\Psi}(\boldsymbol{u}(t))$.

The differences are quite minute between either of the above, thus, for brevity, we choose the latter. In summary, we have the following:

$$\boldsymbol{u}(t+\Delta t) = \boldsymbol{u}(t) + \boldsymbol{v}(t)\Delta t + \frac{\phi(\Delta t)^2}{m}\left(\phi\boldsymbol{\Psi}(\boldsymbol{u}(t + \Delta t)) + (1 - \phi)\boldsymbol{\Psi}(\boldsymbol{u}(t))\right) + \hat{\mathcal{O}}((\Delta t)^2). \quad (8.61)$$

As we have noted earlier:

- When $\phi = 1$, then this is the (implicit) Backward Euler scheme, which is very stable (very dissipative) and $\mathcal{O}((\Delta t)^2)$ locally in time,
- When $\phi = 0$, then this is the (explicit) Forward Euler scheme, which is conditionally stable and $\mathcal{O}((\Delta t)^2)$ locally in time,
- When $\phi = 0.5$, then this is the (implicit) "Midpoint" scheme, which is stable and $\hat{\mathcal{O}}((\Delta t)^2) = \mathcal{O}((\Delta t)^3)$ locally in time.

In summary, for the velocity, we have[8]

$$\boldsymbol{v}(t+\Delta t) = \boldsymbol{v}(t) + \frac{\Delta t}{m}\left(\phi\boldsymbol{\Psi}(t+\Delta t) + (1-\phi)\boldsymbol{\Psi}(t)\right), \tag{8.62}$$

and for the position

$$\begin{aligned}\boldsymbol{u}(t+\Delta t) &= \boldsymbol{u}(t) + \boldsymbol{v}(t+\phi\Delta t)\Delta t \\ &= \boldsymbol{u}(t) + \left(\phi\boldsymbol{v}(t+\Delta t) + (1-\phi)\boldsymbol{v}(t)\right)\Delta t,\end{aligned} \tag{8.63}$$

or more explicitly

$$\boldsymbol{u}(t+\Delta t) = \boldsymbol{u}(t) + \boldsymbol{v}(t)\Delta t + \frac{\phi(\Delta t)^2}{m}\left(\phi\boldsymbol{\Psi}(t+\Delta t) + (1-\phi)\boldsymbol{\Psi}(t)\right). \tag{8.64}$$

In the continuum formulation, $\nabla_x \cdot \boldsymbol{T} + \boldsymbol{f} = \rho\ddot{\boldsymbol{u}}$, $m = \rho$ and $\boldsymbol{\Psi} = \nabla_x \cdot \boldsymbol{T} + \boldsymbol{f}$, where we must apply the (iterative) process introduced earlier to each node in the system. Under infinitesimal deformations, $\nabla_X \cdot \boldsymbol{T} + \boldsymbol{f} = \rho_o \frac{\partial^2 \boldsymbol{u}}{\partial t^2}$, $m = \rho_o$ and $\boldsymbol{\Psi} = \nabla_X \cdot \boldsymbol{T} + \boldsymbol{f}$.

8.5 Spatial Discretization: Spatial Finite Difference Stencils

The following approximations are used:

1. For the first derivative of a primal variable u at (x_1, x_2, x_3):

$$\frac{\partial u}{\partial x_1} \approx \frac{u(x_1+\Delta x_1, x_2, x_3) - u(x_1-\Delta x_1, x_2, x_3)}{2\Delta x_1}. \tag{8.65}$$

2. For the derivative of a flux at (x_1, x_2, x_3):

$$\begin{aligned}\frac{\partial}{\partial x_1}\left(A\frac{\partial u}{\partial x_1}\right) &\approx \frac{\left(A\frac{\partial u}{\partial x_1}\right)|_{x_1+\frac{\Delta x_1}{2}, x_2, x_3} - \left(A\frac{\partial u}{\partial x_1}\right)|_{x_1-\frac{\Delta x_1}{2}, x_2, x_3}}{\Delta x_1} \\ &= \frac{1}{\Delta x_1}\left[A(x_1+\frac{\Delta x_1}{2}, x_2, x_3)\left(\frac{u(x_1+\Delta x_1), x_2, x_3 - u(x_1, x_2, x_3)}{\Delta x_1}\right)\right] \\ &- \frac{1}{\Delta x_1}\left[A(x_1-\frac{\Delta x_1}{2}, x_2, x_3)\left(\frac{u(x_1, x_2, x_3) - u(x_1-\Delta x_1, x_2, x_3)}{\Delta x_1}\right)\right],\end{aligned} \tag{8.66}$$

[8] In order to streamline the notation, we drop the cumbersome $\mathcal{O}(\Delta t)$-type terms.

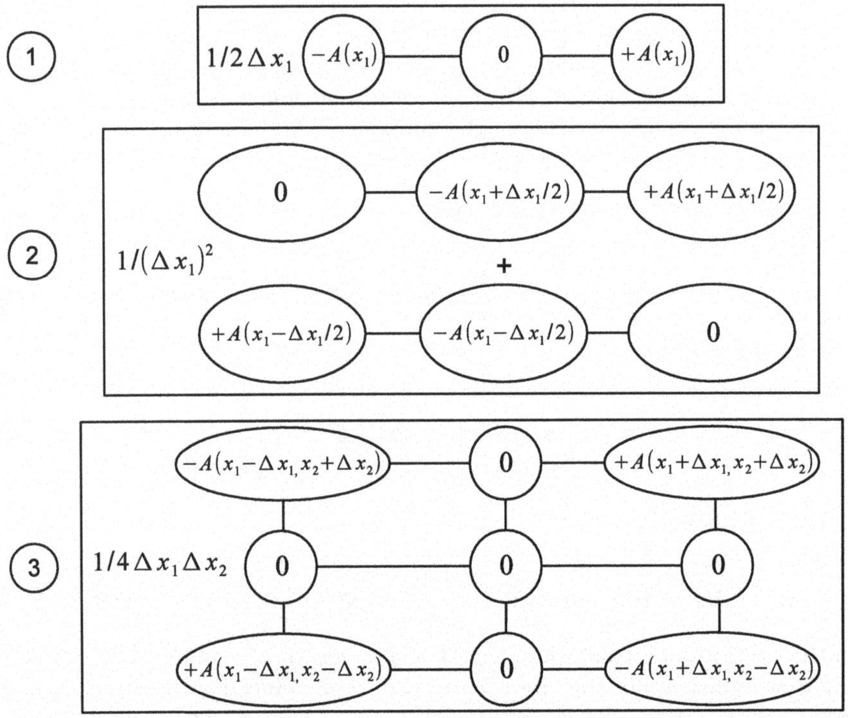

Fig. 8.1. The various stencils in "computational molecule" form (centered at (x_1, x_2, x_3)), where: (1) $\frac{\partial u}{\partial x_1}$, (2) $\frac{\partial}{\partial x_1}\left(A\frac{\partial u}{\partial x_1}\right)$ and (3) $\frac{\partial}{\partial x_2}\left(A\frac{\partial u}{\partial x_1}\right)$ (Zohdi [139]).

where we have used

$$A(x_1 + \frac{\Delta x_1}{2}, x_2, x_3) \approx \frac{1}{2}\left(A(x_1 + \Delta x_1, x_2, x_3) + A(x_1, x_2, x_3)\right) \quad (8.67)$$

and

$$A(x_1 - \frac{\Delta x_1}{2}, x_2, x_3) \approx \frac{1}{2}\left(A(x_1, x_2, x_3) + A(x_1 - \Delta x_1, x_2, x_3)\right). \quad (8.68)$$

3. For the cross-derivative of a flux at (x_1, x_2):

$$\frac{\partial}{\partial x_2}\left(A\frac{\partial u}{\partial x_1}\right) \approx \frac{\partial}{\partial x_2}\left(A(x_1,x_2,x_3)\left(\frac{u(x_1+\Delta x_1,x_2,x_3)-u(x_1-\Delta x_1,x_2,x_3)}{2\Delta x_1}\right)\right)$$

$$\approx \frac{1}{4\Delta x_1 \Delta x_2}(A(x_1,x_2+\Delta x_2,x_3)\left[u(x_1+\Delta x_1,x_2+\Delta x_2,x_3)-u(x_1-\Delta x_1,x_2+\Delta x_2,x_3)\right]$$
$$-\ A(x_1,x_2-\Delta x_2,x_3)\left[u(x_1+\Delta x_1,x_2-\Delta x_2,x_3)-u(x_1-\Delta x_1,x_2-\Delta x_2,x_3)\right]). \qquad (8.69)$$

9

A Model Problem: Dielectrics Undergoing Multifield Processes

9.1 Introduction

An important issue that this monograph addresses is the modeling and simulation of strongly coupled electromagnetic and thermodynamic fields that arise in particulate-doped dielectrics using an adaptive staggered FDTD (Finite Difference Time Domain) method. Of particular interest is to provide a straightforward modular approach to finding the effective dielectric (electromagnetic) response of a material, incorporating thermal effects arising from Joule heating which alter the pointwise dielectric properties such as the electric permittivity, magnetic permeability, and electric conductivity. This is important for "thermal (damage) management" of materials used in electromagnetic applications. Because multiple field coupling is present, a staggered, temporally-adaptive scheme is developed to resolve the internal microstructural electric, magnetic and thermal fields, accounting for the simultaneous pointwise changes in the material properties. Numerical examples are provided to illustrate the approach. Extensions to coupled chemical and mechanical fields are also provided. Here, we follow an approach found in Zohdi [139].

9.1.1 Thermal Coupling

In general, the properties of most electromagnetic materials are quite sensitive to the temperature. For example, let us consider the three main electromagnetic properties, the electrical permittivity, magnetic permeability and electrical conductivity, in the isotropic case ($\boldsymbol{\epsilon} = \epsilon \mathbf{1}$, $\boldsymbol{\mu} = \mu \mathbf{1}$ and $\boldsymbol{\sigma} = \sigma \mathbf{1}$). Following common practice, for electromagnetic materials, we write $\epsilon = \epsilon_o \epsilon_r$ where $\epsilon_o = 8.854 \times 10^{-12}$ farads/meter is the free space permittivity and ϵ_r is the relative permittivity or "dielectric" constant. Also, we have $\mu_o = 4\pi \times 10^{-7} Ns^2/C^2$ and $\mu = \mu_r \mu_o$, where μ_r is the relative magnetic permeability. As a "model problem" material, we consider the following

T.I. Zohdi: Electromagnetic Properties of Multiphase Dielectrics, LNACM 64, pp. 103–150.
springerlink.com

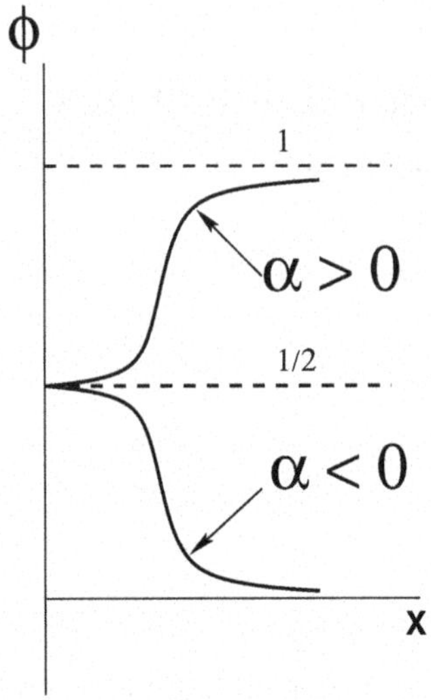

Fig. 9.1. General trends for Sigmoid-type behavior in which $\phi = \left(1 + e^{-\alpha x}\right)^{-1}$ (Zohdi [139]).

decompositions, employing thermoelectromagnetic saturation conditions (Sigmoid functions, Figure 9.1):

- For the electrical permittivity :

$$\epsilon(\theta, \boldsymbol{E}) = \epsilon_o(1+\chi_E(\theta, \boldsymbol{E})) = \epsilon_o\epsilon_r(\theta, \boldsymbol{E}) = \epsilon_o\epsilon_r(\theta_R, \boldsymbol{E}_R)\mathcal{E}(\theta-\theta_R, \boldsymbol{E}-\boldsymbol{E}_R), \tag{9.1}$$

  where χ_E is the electric susceptibility, θ is the temperature and the last term is a representation around a reference state, for example, using saturation-type Sigmoid functions of the form

$$\mathcal{E}(\theta-\theta_R, \boldsymbol{E}-\boldsymbol{E}_R) = 1 + \mathcal{K}_{E1}\left(1+e^{-\alpha_{E1}(\theta-\theta_R)}\right)^{-1} + \mathcal{K}_{E2}\left(1+e^{-\alpha_{E2}||\boldsymbol{E}-\boldsymbol{E}_R||}\right)^{-1}, \tag{9.2}$$

  where the α's and $\mathcal{K}$'s are material parameters, and the terms with subscript "R" are reference values.
- For the magnetic permeability :

$$\mu(\theta, \boldsymbol{E}) = \mu_o(1+\chi_H(\theta, \boldsymbol{H})) = \mu_o\mu_r(\theta, \boldsymbol{H}) = \mu_o\mu_r(\theta_R, \boldsymbol{H}_R)\mathcal{H}(\theta-\theta_R, \boldsymbol{H}-\boldsymbol{H}_R), \tag{9.3}$$

  where χ_H is the magnetic susceptibility and, consistent with the previous electrical decomposition

$$\mathcal{H}(\theta - \theta_R, \boldsymbol{H} - \boldsymbol{H}_R) = 1 + \mathcal{K}_{H1}\left(1 + e^{-\alpha_{H1}(\theta-\theta_R)}\right)^{-1} + \mathcal{K}_{H2}\left(1 + e^{-\alpha_{H2}||\boldsymbol{H}-\boldsymbol{H}_R||}\right)^{-1}. \tag{9.4}$$

- For the electrical conductivity:

$$\sigma(\theta, \boldsymbol{E}) = \sigma(\theta_R, \boldsymbol{E}_R)\left(1 + \mathcal{K}_{S1}(1 + e^{-\alpha_{S1}(\theta-\theta_R)})^{-1} + \mathcal{K}_{S2}(1 + e^{-\alpha_{S2}||\boldsymbol{E}-\boldsymbol{E}_R||})^{-1}\right). \tag{9.5}$$

Generally speaking, for many materials, until a saturation threshold is met, $\epsilon_r(\theta, \boldsymbol{E})$ grows with θ, $\mu_r(\theta, \boldsymbol{H})$ decreases with θ and $\sigma(\theta)$ decreases with θ. Because the electromagnetic field and subsequent flow of current through real materials leads to Joule heating, producing changes in the pointwise material properties, analytical predictions are somewhat limited, and one must resort to numerical schemes and so-called "mesoscale" computation, posed over a statistically representative volume element sample containing significant microstructure (several particles).

Remark: See the treatise of Jackson [46] for reviews of the rich variety of possible dielectrical responses of materials, including atomistic-level discussions to motivate nonlinear dielectric behavior. Later, we shall utilize the previously mentioned specific "model-problem" decompositions (Equations 9.1-9.5), however, the numerical formulations are developed for general cases.

9.1.2 Numerical Methods

In order to properly capture the coupled (transient) electromagnetic and thermal behavior of a new material, Maxwell's equations, coupled to the first law of thermodynamics, must be solved (simultaneously) over a representative volume element. Thereafter, the overall thermally-sensitive properties can be directly postprocessed via volumetric averaging. The primary goal of this work is to develop numerical methods in order to ascertain the effects of particulates on the overall coupled response of thermally-sensitive dielectrics. There are a variety of Computational Electromagnetic Methods (CEM) which can be lumped into two broad categories: Differential Equation Formulations (DEF) and Integral Formulations (IF). In the DEF category, the most widely used techniques is the *Finite Difference Time Domain Method* (FDTD), which is ideally suited to the problems of interest in this work. FDTD will serve as a foundation of the approach developed in this work. However, in passing, we will mention other DEF-based methods, such as,

- *Multiresolution time domain method:* solution techniques based on wavelet-based discretization,

- *Finite element method:* Solution techniques based on discretization of variational formulations and which are ideal for irregular geometries,[1]
- *Pseudospectral time domain method:* Solution techniques based on Fourier and Chebyshev transforms, followed by a lattice or grid discretization of the transformed domain,

while in the IF category one has, for example,

- *Discrete dipole approximation:* Solution techniques based on an array of dipoles solved iteratively with the conjugate gradient method and a fast Fourier transform to multiply matrices,
- *Method of moments:* Solution techniques based on integral formulations employing boundary element method discretization, often accompanied by the fast multipole method to accelerate summations needed during the calculations,
- *Partial element equivalent circuit method:* Solution techniques based on integral equations that are interpreted as circuits in discretization cells. This approach is ideal for circuit layout design.

The presentation is divided into three main parts: (1) formulations for each field, identifying the coupling terms, (2) iterative staggering schemes (including spatial and temporal discretization) and (3) numerical examples for model problems. Initially, we will focus on the electromagneto-thermo fields, and then extend the formulations to include chemical and mechanical fields. This study utilizes and extends low order (first order in space and time) FEM-based techniques originally developed in Zohdi [129, 131] to simulate longterm diffusive systems modeling corrosion and to develop second order (in both space and time) FDTD methods for electro-magneto-thermo-chemo-mechano-type systems of interest to the electronics industry. The approach builds on work found in Zohdi [137, 139], enlarging the types of fields in the system to include electro-magneto-thermo-chemo-mechano-type systems in addition to types of (semiconductor-like) constitutive relations for the electric permittivity , magnetic permeability and electrical conductivity, which are nonmonotone Sigmoid-type constitutive of laws and providing a more in-depth analysis of this type of system behavior. The developed approach is implicit, second order accurate and relatively easy to encode. Unlike usual FDTD (explicit) methods, the formulation developed in this work is implicit, and is relatively robust, employing an adaptive staggering scheme to capture evolving multi-physics. This formulation is important for material designers who seek ways by which to modify a base material, for example, by employing particulate additives, however, simultaneously avoiding inadvertent overheating and thermal stresses.

[1] In particular, see Demkowicz et al. [14, 15] and Rachowicz and Zdunek [89] for the state of the art in adaptive finite element methods for harmonic Maxwell's equations.

9.2 Transient Electro-Magneto-Thermo Coupled Fields

We now provide the essential field equations that will be used during the mesoscale computation.

9.2.1 Electromagnetic Fields: Maxwell's Equations

In order to generate the overall (volume averaged) thermoelectro-magnetic response of a heterogeneous continuum sample, we solve Maxwell's equations posed over a representative volume element (RVE) domain by starting with Faraday's law

$$\nabla_x \times \boldsymbol{E} = -\left(\frac{\partial \boldsymbol{B}}{\partial t} + \boldsymbol{M}^{ext} + \hat{\boldsymbol{\sigma}} \cdot \boldsymbol{H}\right) \tag{9.6}$$

and Ampere's law

$$\nabla_x \times \boldsymbol{H} = \frac{\partial \boldsymbol{D}}{\partial t} + \boldsymbol{J}^{ext} + \boldsymbol{\sigma} \cdot \boldsymbol{E}, \tag{9.7}$$

where we recall that $\boldsymbol{E}$ is the electric field, $\boldsymbol{D} = \boldsymbol{\epsilon} \cdot \boldsymbol{E}$ is the electric field flux, $\boldsymbol{J}^{ext}$ is the source electric current, $\boldsymbol{H}$ is the magnetic field, $\boldsymbol{B} = \boldsymbol{\mu} \cdot \boldsymbol{H}$ is the magnetic field flux, $\boldsymbol{M}^{ext}$ is the source "equivalent magnetic current," $\boldsymbol{\epsilon}$ is the electric permittivity, $\boldsymbol{\mu}$ is the magnetic permeability, $\boldsymbol{\sigma}$ is the electric conductivity and $\hat{\boldsymbol{\sigma}}$ is the equivalent magnetic loss. The material is assumed to be heterogeneous (spatially variable), isotropic, and thermally-sensitive.

Remark: $\boldsymbol{M}^{ext}$ is a phenomenological term that frequently appears in the literature to account for magnetic "sources/losses" and "magnetic conduction." We shall keep these terms throughout the formulations, but with an implicit "warning" that they are difficult to justify from first principles. Furthermore, $\hat{\boldsymbol{\sigma}}$ can be considered as the equivalent *phenomenological* magnetic "conductivity" (loss).

9.2.2 Thermodynamics: First Law and Absorption of Energy

The interconversions of various forms of energy (electromagnetic, thermal, etc.) in a system are governed by the first law of thermodynamics,

$$\rho \dot{w} - \boldsymbol{T} : \nabla_x \dot{\boldsymbol{u}} + \nabla_x \cdot \boldsymbol{q} - \rho z = 0, \tag{9.8}$$

where w is the stored energy per unit mass (which is a function of the temperature, θ), ρ is the density, $\boldsymbol{T}$ is Cauchy stress, $\boldsymbol{u}$ is the displacement field, $\boldsymbol{q}$ is heat flux, and ρz is the rate of electromagnetic energy absorbed due to Joule heating,

$$\rho z = a\left(\boldsymbol{E} \cdot \boldsymbol{J}^{tot} + \boldsymbol{H} \cdot \boldsymbol{M}^{tot}\right), \tag{9.9}$$

where $0 \leq a \leq 1$ is an absorption constant, $\boldsymbol{J}^{tot} \stackrel{\text{def}}{=} \boldsymbol{J}^{ext} + \boldsymbol{\sigma} \cdot \boldsymbol{E}$ and $\boldsymbol{M}^{tot} \stackrel{\text{def}}{=} \boldsymbol{M}^{ext} + \hat{\boldsymbol{\sigma}} \cdot \boldsymbol{H}$. If we consider the effects of deformation and stress to be insignificant,

$$\rho \dot{w} + \nabla_x \cdot \boldsymbol{q} - \rho z = 0, \tag{9.10}$$

and consider the stored thermal energy per unit mass to be $w = C\theta$ and Fourier's $\boldsymbol{q} = -\mathbb{K} \cdot \nabla_x \theta$. These assumptions lead to

$$\rho C \dot{\theta} - \nabla_x \cdot \mathbb{K} \cdot \nabla_x \theta - a \left(\boldsymbol{E} \cdot \boldsymbol{J}^{tot} + \boldsymbol{H} \cdot \boldsymbol{M}^{tot} \right) = 0, \tag{9.11}$$

where C is the heat capacity per unit mass and $\mathbb{K}$ is the thermal conductivity.

Remark 1: As mentioned in the introduction, we consider the pointwise properties, $\boldsymbol{\epsilon}(\boldsymbol{x})$, $\boldsymbol{\mu}(\boldsymbol{x})$ and $\boldsymbol{\sigma}(\boldsymbol{x})$ to be thermally-dependent, for example, governed by Equations 9.1-9.5.

Remark 2: Joule heating terms have been derived earlier.

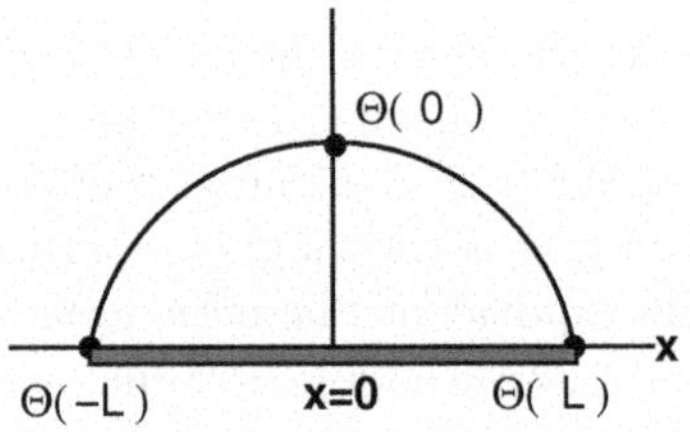

Fig. 9.2. A one-dimensional structure (Zohdi [139]).

Remark 3: For illustration purposes, if one considers a one-dimensional steady-state problem (Figure 9.2), ignoring the effects of stress, and assuming $\rho C \dot{\theta} = \rho C \dot{w}$,

$$\rho C \dot{\theta} = 0 = K \frac{d^2 \theta}{dx^2} + JE \Rightarrow \theta(x) = \theta(L) + \frac{JE}{2K}(L^2 - x^2), \tag{9.12}$$

the maximum temperature becomes ($H = JE$, and assuming $J = \sigma E$ or $E = RJ$), that is,

$$\theta(0) = \theta^{max} = \theta(L) + \underbrace{\frac{RJ^2}{2K} L^2}_{\text{thermal source}}, \tag{9.13}$$

which clearly shows the contribution of the Joule heating to the rise in temperature.

9.3 Numerical Simulations: Staggering Schemes for Thermoelectro-Magnetically Coupled Problems

We now develop a staggering solution framework to solve the coupled systems of interest. The general methodology is as follows (at a given time increment): (1) each field equation is solved individually, "freezing" the other (coupled) fields in the system, only allowing the primary field to be active and (2) after the solution of each field equation, the primary field variable is updated, and the next field equation is treated in a similar manner. For an "implicit" type of staggering, the process can be repeated in an iterative manner, while for an "explicit" type, one moves to the next time step after one "pass" through the system. We will employ implicit staggering. Specifically, for the thermoelectro-magneto system under consideration, consider an abstract setting, whereby one solves for the electric field, assuming the magnetic and thermal fields are fixed (L is a time step counter and K is a staggering-step counter),

$$\mathcal{A}_1(\underline{\boldsymbol{E}^{L+1,K}}, \boldsymbol{H}^{L+1,K-1}, \theta^{L+1,K-1}) = \mathcal{B}_1(\boldsymbol{E}^{L+1,K-1}, \boldsymbol{H}^{L+1,K-1}, \theta^{L+1,K-1}). \tag{9.14}$$

Then, one solves for the magnetic fields, assuming the electric and thermal fields fixed,

$$\mathcal{A}_2(\boldsymbol{E}^{L+1,K}, \underline{\boldsymbol{H}^{L+1,K}}, \theta^{L+1,K-1}) = \mathcal{B}_2(\boldsymbol{E}^{L+1,K}, \boldsymbol{H}^{L+1,K-1}, \theta^{L+1,K-1}). \tag{9.15}$$

Then, one solves for the thermal fields, assuming the electric and magnetic fields fixed,

$$\mathcal{A}_3(\boldsymbol{E}^{L+1,K}, \boldsymbol{H}^{L+1,K}, \underline{\theta^{L+1,K}}) = \mathcal{B}_3(\boldsymbol{E}^{L+1,K}, \boldsymbol{H}^{L+1,K}, \theta^{L+1,K-1}), \tag{9.16}$$

where the only underlined variable is "active" at that stage of the process. Within the staggering (iterative) scheme, implicit time-stepping methods (with time step size adaptivity) will be used throughout the following analysis. The process is driven by minimizing the following normalized errors (for the three fields) within each time step, which represent the nondimensional ratios for each field of the iterative error within a time step (difference between successive iterations) to the difference in the converged solution from time step to time step; for the electric field and determine whether the the iterations should continue, or the time steps should be adaptively reduced (or increased if convergence occurs too quickly). The details of this process are discussed shortly. Generally speaking, if a recursive staggering process is not employed (an explicit coupling scheme), the staggering error can accumulate rapidly. However, simply employing extremely small time steps, smaller than needed to control the discretization error, in order to suppress a (nonrecursive) staggering process error, can be computationally inefficient. Therefore, the objective of the next subsection is to develop a strategy to adaptively adjust, in fact maximize, the choice of the time step size in order to control the staggering error, while simultaneously staying below a critical time step size needed to control the discretization error. An important related issue is to simultaneously minimize the computational effort involved. We now develop a staggering scheme by extending an approach found in the work of Zohdi et al. [129–134 and 137, 139].

Remark 1: The symbol $||\cdot||$ will signify the following metric for the electric field

$$\varpi_E^K \stackrel{\text{def}}{=} \frac{||\boldsymbol{E}^{L+1,K} - \boldsymbol{E}^{L+1,K-1}||}{||\boldsymbol{E}^{L+1,K} - \boldsymbol{E}^{L}||}, \tag{9.17}$$

and for the magnetic field

$$\varpi_H^K \stackrel{\text{def}}{=} \frac{||\boldsymbol{H}^{L+1,K} - \boldsymbol{H}^{L+1,K-1}||}{||\boldsymbol{H}^{L+1,K} - \boldsymbol{H}^{L}||}, \tag{9.18}$$

and for the thermodynamic field

$$\varpi K_\theta \stackrel{\text{def}}{=} \frac{||\theta^{L+1,K} - \theta^{L+1,K-1}||}{||\theta^{L+1,K} - \theta^{L}||}. \tag{9.19}$$

Thereafter, we select the maximum *nondimensionalized* error for adaptivity

$$\varpi^{*,K} \stackrel{\text{def}}{=} max(\varpi_E^K, \varpi_H^K, \varpi_\theta^K). \tag{9.20}$$

Remark 2: Staggering schemes are widely used in the computational mechanics literature, dating back, at least, to Zienkiewicz [112] and Zienkiewicz et al. [113]. For in-depth overviews, see the works of Lewis and Schrefler (Lewis et al. [57] and Lewis and Schrefler [58]), and a series of works by Schrefler and collaborators: Schrefler [92], Turska and Schrefler [105], Bianco et al. [8] and Wang and Schrefler [107]. For a broad cross-sectional view of the related literature, we refer the reader to: Doltsinis [16] and [17], Farhat et al. [24], Farhat and Lesoinne [25], Farhat et al. [26], Fish and Chen [27], Lesoinne and Farhat [55], Le Tallec and Mouro [56], Michopoulos et al. [71], Park and Felippa [77], Piperno [78], Piperno et al. [87], Piperno and Farhat [88].

9.3.1 Spatial Discretization of the Coupled System

Spatial discretization of the electromagnetic field

Numerically, the components of the curl of functions such as $\boldsymbol{E}$ and $\boldsymbol{H}$ are approximated by central finite difference stencils of the form (Figure 9.3):

$$\frac{\partial \boldsymbol{E}}{\partial x}|_x \approx \frac{\boldsymbol{E}(x+\Delta x) - \boldsymbol{E}(x-\Delta x)}{2\Delta x} \text{ and } \frac{\partial \boldsymbol{H}}{\partial x}|_x \approx \frac{\boldsymbol{H}(x+\Delta x) - \boldsymbol{H}(x-\Delta x)}{2\Delta x}, \tag{9.21}$$

for each of the (x_1, x_2, x_3)-directions, in order to form the terms needed in $\nabla_x \times \boldsymbol{E}$ and $\nabla_x \times \boldsymbol{H}$. This is a second order accurate stencil. To illustrate this, consider a Taylor series expansion for an arbitrary function A, that is,

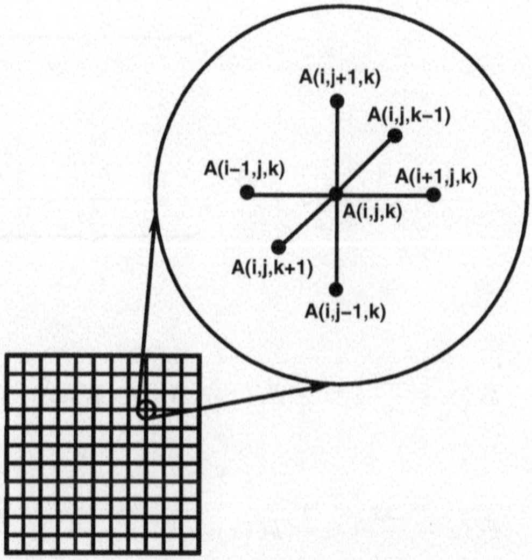

Fig. 9.3. A typical three-dimensional finite-difference stencil for a field $A(x, y, z)$ (Zohdi [139]).

$$A(x + \Delta x) = A(x) + \frac{\partial A}{\partial x}|_x \Delta x + \frac{1}{2}\frac{\partial^2 A}{\partial x^2}|_x (\Delta x)^2 + \frac{1}{6}\frac{\partial^3 A}{\partial x^3}|_x (\Delta x)^3 + \mathcal{O}((\Delta x)^4) \tag{9.22}$$

and

$$A(x - \Delta x) = A(x) - \frac{\partial A}{\partial x}|_x \Delta x + \frac{1}{2}\frac{\partial^2 A}{\partial x^2}|_x (\Delta x)^2 - \frac{1}{6}\frac{\partial^3 A}{\partial x^3}|_x (\Delta x)^3 + \mathcal{O}((\Delta x)^4). \tag{9.23}$$

Subtracting the two expressions yields

$$\frac{\partial A}{\partial x}|_x = \frac{A(x + \Delta x) - A(x - \Delta x)}{2\Delta x} + \mathcal{O}((\Delta x)^2). \tag{9.24}$$

Spatial discretization of the thermal field

Although the discretization of the thermal field follows the same approach as the electromagnetic field, because it is governed by a second order differential equation, the following approximation is first made, that is,

$$\frac{\partial q}{\partial x}|_x \approx \frac{q(x + \frac{\Delta x}{2}) - q(x - \frac{\Delta x}{2})}{\Delta x}, \tag{9.25}$$

where (in conjunction with Fourier's law)

$$q(x+\frac{\Delta x}{2}) \approx -\mathbb{K}(x+\frac{\Delta x}{2})\underbrace{\frac{\theta(x+\Delta x)-\theta(x)}{\Delta x}}_{\frac{\partial\theta}{\partial x}|_{x+\frac{\Delta x}{2}}} \tag{9.26}$$

and

$$q(x-\frac{\Delta x}{2}) \approx -\mathbb{K}(x+\frac{\Delta x}{2})\underbrace{\frac{\theta(x)-\theta(x-\Delta x)}{\Delta x}}_{\frac{\partial\theta}{\partial x}|_{x-\frac{\Delta x}{2}}} \tag{9.27}$$

where

$$\mathbb{K}(x+\frac{\Delta x}{2}) \approx \frac{1}{2}(\mathbb{K}(x+\Delta x)+\mathbb{K}(x)), \tag{9.28}$$

and

$$\mathbb{K}(x-\frac{\Delta x}{2}) \approx \frac{1}{2}(\mathbb{K}(x)+\mathbb{K}(x-\Delta x)). \tag{9.29}$$

These approximations are made for $\frac{\partial q_1}{\partial x_1}$, $\frac{\partial q_2}{\partial x_2}$ and $\frac{\partial q_3}{\partial x_3}$ in order to form the terms needed in $\nabla_x \cdot \boldsymbol{q}$.

9.3.2 Temporal Discretization of the Coupled System

Temporal discretization of the electromagnetic fields

We start with a relatively general, "lossy," formulation of Maxwell's equations (Faraday's law and Ampere's law),

$$\frac{\partial(\boldsymbol{\mu}\cdot\boldsymbol{H})}{\partial t} = -\nabla_x \times \boldsymbol{E} - \boldsymbol{M}^{ext} - \hat{\boldsymbol{\sigma}}\cdot\boldsymbol{H} \stackrel{\text{def}}{=} \boldsymbol{F} \tag{9.30}$$

and

$$\frac{\partial(\boldsymbol{\epsilon}\cdot\boldsymbol{E})}{\partial t} = \nabla_x \times \boldsymbol{H} - \boldsymbol{J}^{ext} - \boldsymbol{\sigma}\cdot\boldsymbol{E} \stackrel{\text{def}}{=} \boldsymbol{G}. \tag{9.31}$$

We discretize for time$=t+\phi\Delta t$, and using a trapezoidal "$\phi - scheme$" $(0 \leq \phi \leq 1)$,

$$\frac{(\boldsymbol{\mu}\cdot\boldsymbol{H})(t+\Delta t)-(\boldsymbol{\mu}\cdot\boldsymbol{H})(t)}{\Delta t} \approx \boldsymbol{F}(t+\phi\Delta t) \approx \phi\boldsymbol{F}(t+\Delta t)+(1-\phi)\boldsymbol{F}(t) \tag{9.32}$$

and

$$\frac{(\boldsymbol{\epsilon}\cdot\boldsymbol{E})(t+\Delta t)-(\boldsymbol{\epsilon}\cdot\boldsymbol{E})(t)}{\Delta t} \approx \boldsymbol{G}(t+\phi\Delta t) \approx \phi\boldsymbol{G}(t+\Delta t)+(1-\phi)\boldsymbol{G}(t). \tag{9.33}$$

Rearranging yields

$$\boldsymbol{H}(t+\Delta t) \approx \boldsymbol{\mu}^{-1}(t+\Delta t)\cdot((\boldsymbol{\mu}\cdot\boldsymbol{H})(t)+\Delta t\,(\phi\boldsymbol{F}(t+\Delta t)+(1-\phi)\boldsymbol{F}(t))) \tag{9.34}$$

and

$$\boldsymbol{E}(t+\Delta t) \approx \boldsymbol{\epsilon}^{-1}(t+\Delta t)\cdot((\boldsymbol{\epsilon}\cdot\boldsymbol{E})(t)+\Delta t\,(\phi\boldsymbol{G}(t+\Delta t)+(1-\phi)\boldsymbol{G}(t)))\,, \tag{9.35}$$

where the previously introduced spatial discretization is applied to the terms in $\boldsymbol{F}$ and $\boldsymbol{G}$ ($\nabla_x \times \boldsymbol{H}$ and $\nabla_x \times \boldsymbol{E}$). Clearly, other constitutive laws, for example, for the conduction, $\boldsymbol{J} = \mathcal{J}(\boldsymbol{E})$, can easily be handled with the same formulation. Note that the nonlinearity can be far more complicated and not easily separable, and one must resort to using a Taylor series expansion of the form

$$\boldsymbol{D} = f(\boldsymbol{E}^K) \approx f(\boldsymbol{E}^{K-1}) + \frac{\partial f}{\partial \boldsymbol{E}}|_{\boldsymbol{E}^{K-1}} \cdot (\boldsymbol{E}^K - \boldsymbol{E}^{K-1}) + ... + \mathcal{O}((\boldsymbol{E})^2) \tag{9.36}$$

and

$$\boldsymbol{B} = g(\boldsymbol{H}^K) \approx g(\boldsymbol{H}^{K-1}) + \frac{\partial g}{\partial \boldsymbol{H}}|_{\boldsymbol{H}^{K-1}} \cdot (\boldsymbol{H}^K - \boldsymbol{H}^{K-1}) + ... + \mathcal{O}((\boldsymbol{H})^2) \tag{9.37}$$

to linearize the system. Depending on the type of nonlinearity, there may be slight advantages in arranging the fixed point iteration in a certain order over another.

Temporal discretization of the thermal field

For the thermal field, we write

$$\frac{\partial \theta}{\partial t} = \frac{1}{\rho C}(-\nabla_x \cdot \boldsymbol{q} + \rho z) \stackrel{\text{def}}{=} Y. \tag{9.38}$$

We discretize around the time$=t+\phi\Delta t$, yielding

$$\theta(t+\Delta t) \approx \theta(t) + \Delta t\,(\phi Y(t+\Delta t) + (1-\phi)Y(t))\,, \tag{9.39}$$

where the previously introduced spatial discretization is applied to the terms in Y.

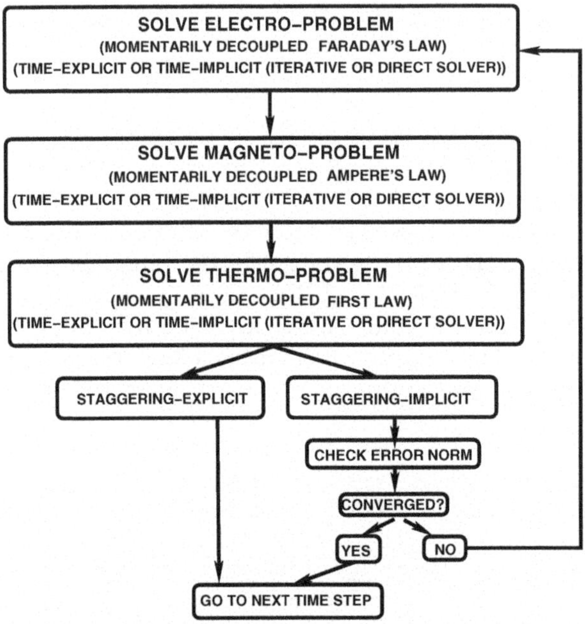

Fig. 9.4. Types of coupled staggering solution for the thermoelectro-magneto system (Zohdi [139]).

9.3.3 The Overall Solution Scheme

In order to construct a solution, the algorithm is as follows:

- (1) *Spatiotemporal discretization*: Construct derivative terms such as

$$\frac{\partial \boldsymbol{E}(x)}{\partial x} \approx \frac{\boldsymbol{E}(x+\Delta x) - \boldsymbol{E}(x-\Delta x)}{2\Delta x}, \; etc. \tag{9.40}$$

  and insert into the governing equations. This leads to a system of coupled equations, for each node ((i,j,k) in Figure 9.3), which are cast in the following (implicit/recursive) form, that is,

$$\boldsymbol{E}(t+\Delta t) = \mathcal{F}(\boldsymbol{E}(t+\Delta t), \boldsymbol{H}(t+\Delta t), \theta(t+\Delta t), ...), \tag{9.41}$$

  and

$$\boldsymbol{H}(t+\Delta t) = \mathcal{G}(\boldsymbol{E}(t+\Delta t), \boldsymbol{H}(t+\Delta t), \theta(t+\Delta t), ...), \tag{9.42}$$

and

$$\theta(t + \Delta t) = \mathcal{Y}(\boldsymbol{E}(t + \Delta t), \boldsymbol{H}(t + \Delta t), \theta(t + \Delta t), ...). \quad (9.43)$$

- (2) *System staggering:* Compute $\boldsymbol{E}$-field with $\boldsymbol{H}$ and θ fields fixed, then compute $\boldsymbol{H}$-field with $\boldsymbol{E}$ and θ fields fixed, etc., and iterate at time interval $L+1$, $K = 1, 2, ...$ for

$$\boldsymbol{E}^{L+1,K} = \mathcal{F}(\underline{\boldsymbol{E}^{L+1,K-1}}, \boldsymbol{H}^{L+1,K-1}, \theta^{L+1,K-1}), \quad (9.44)$$

$$\boldsymbol{H}^{L+1,K} = \mathcal{G}(\boldsymbol{E}^{L+1,K}, \underline{\boldsymbol{H}^{L+1,K-1}}, \theta^{L+1,K-1}), \quad (9.45)$$

and

$$\theta^{L+1,K} = \mathcal{Y}(\boldsymbol{E}^{L+1,K}, \boldsymbol{H}^{L+1,K}, \underline{\theta^{L+1,K-1}}). \quad (9.46)$$

Solving each of the above equations (9.44, 9.45, 9.46), with the respective other fields fixed can be achieved in a variety of ways, for example, iteratively or by direct (Gaussian-type) solution methods (Figure 9.4). However, in theory, one could even simply perform an explicit update (no recursion). This is discussed further in the remarks that follow.

- (3) Compute error measures: $\varpi^{*,K} \stackrel{\text{def}}{=} max(\varpi_E^K, \varpi_H^K, \varpi_\theta^K)$.
- (4a) If tolerance is met, $\varpi^{*,K} \leq C_{tol}$ and $K \leq K_d$, then:
  (i) Increment time forward: $t = t + \Delta t$,
  (ii) Construct new time step: $(\Delta t)^{new} = \Phi_K (\Delta t)^{\text{old}}$, where

$$\Phi_K \stackrel{\text{def}}{=} \left(\frac{(\frac{C_{tol}}{\varpi*,0})^{\frac{1}{pK_d}}}{(\frac{\varpi^{*,K}}{\varpi^{*,0}})^{\frac{1}{pK}}} \right). \quad (9.47)$$

  (iii) Select $\Delta t = min((\Delta t)^{lim}, (\Delta t)^{new})$ and go to (1)
- (4b) If tolerance is not met, $\varpi^{*,K} > C_{tol}$ and $K = K_d$, then construct (refine) new time step: $(\Delta t)^{new} \stackrel{\text{def}}{=} \Phi_K (\Delta t)^{\text{old}}$

$$\Phi_K \stackrel{\text{def}}{=} \left(\frac{(\frac{C_{tol}}{\varpi*,0})^{\frac{1}{pK_d}}}{(\frac{\varpi^{*,K}}{\varpi^{*,0}})^{\frac{1}{pK}}} \right) \quad (9.48)$$

  and go to (1). This adaptive time-scaling law (Zohdi [129–134, 137,139]) was derived earlier in the monograph.

At a given time, once the process is complete, then the time is incremented forward and the process is repeated. The overall goal is to deliver solutions where the iterative error is controlled and the temporal discretization accuracy dictates the upper limit on the time step size (Δt^{lim}). Clearly, there are various combinations of solution methods that one can choose from. For example, for the overall field coupling, one may choose implicit or explicit staggering , and within the staggering process, either implicit ($0 < \phi \leq 1$) or explicit time-stepping ($\phi = 0$), and in the case of implicit time-stepping, iterative or direct solvers for Maxwell's equations and the first law of thermodynamics (Figure 9.4).

9.3.4 Discussion of the Numerical Scheme

There are a series of observations pertaining to the numerical scheme.

Comments on the staggering approach

Generally speaking, the solution to the individual field equations progresses in a node by node fashion, whereby at a node (i, j, k), for example, for the magnetic field calculations, one has

$$\boldsymbol{H}(t+\Delta t)_{i,j,k} \approx \boldsymbol{\mu}^{-1}(t+\Delta t)_{i,j,k} \cdot \Big((\boldsymbol{\mu} \cdot \boldsymbol{H})(t)_{i,j,k} + \Delta t\,(\phi \boldsymbol{F}(t+\Delta t) + (1-\phi)\boldsymbol{F}(t))_{i,j,k}\Big), \tag{9.49}$$

where the term on the left-hand side is updated and the terms on the right are previous iterate (old) values. This entails using the old values for all finite difference stencils that eventually become updated only after the algorithm completely "moves through" the system, updating values, node by node (*no matrices need to be formed*). There are a variety of techniques to accelerate such computations, such as successive overrelaxation, based on the pioneering work of Young [111]. For reviews, see Ames [3] or Axelsson [5]. Note that for the magnetic field calculations (Ampere's law), the electric field and thermal fields are "momentarily frozen," and are updated only when it is the appropriate field's "turn" to be solved in the staggered manner. At the algebraic equation solution level, after the individual field has been solved, the entire solution is passed to the next field equation, as described in the previous algorithm (Figure 9.4). This is a recursive iterative scheme of the Jacobi-type, whereby the updates are made only after one complete system iteration (as illustrated in the previous algorithm). The Jacobi method is easier to address theoretically, while the Gauss–Seidel method, which involves immediately using the most current field values when they become available, is usually used at the implementation level. It is important to realize that the Jacobi method is easily parallelizable. In other words, the calculation for each node is (momentarily) uncoupled, with the updates only coming at the end of an iteration. Gauss–Seidel, since it requires the most current updates, couples the nodal calculations immediately. However, these methods can be combined to create hybrid "block-partitioned" approaches, whereby the entire domain is partitioned into subdomains and within each subdomain a Gauss–Seidel method is applied. In other words, for a subdomain, the values at all nodes from the outside are initially frozen, as far as calculations involving members of the group are concerned ("block"-type calculations; see Axelsson [5]). After each isolated subdomain's solution (nodal values) has converged (computed in parallel), then all nodal values are updated, i.e., the most current values become available to all members of the grid, and the isolated subdomain calculations are repeated. Although parallel computations is not the subject of this work, some comments are also in order. Generally, interprocessor communication and synchronization is the usual bottleneck to obtain a high-performance parallel algorithm. The parallel speedup (relative to a sequential implementation), S, can be approximated by Amdahl's law (Amdahl [2]), $S = \frac{1}{1-P}$, where P is the fraction of the algorithm that is parallelizable. For example, if 40 % of the code is inherently sequential, then $P = 0.6$ and $S = 2.5$. This provides an upper bound on the utility of adding more processors. A related expression is "Gustafson's law" (Gustafson [38]), $S(P) = P - k(P-1)$, where k represents the parts of the algorithm that are not parallelizable. Amdahl's law assumes that the problem is of fixed size and that the sequential part is independent of the number of processors, however, Gustafson's law does not make either of these assumptions. Although we do not pursue parallel computation in the present work, we refer the reader to the works of Papadrakakis et al. [79-81] for parallel strategies that are directly applicable to the class of problems of interest, and to Stavroulakis and Papadrakakis [96] for the cutting edge analysis of block-parallel type calculations.

Comments on time step adaptivity

Clearly, one should use the previous (converged) time step's solution as the starting guess for the next time step to obtain a "head-start" ($\boldsymbol{H}^{K=0}(t+\Delta t) = \boldsymbol{H}(t)$). When selecting a time step, one must balance accuracy concerns and, simultaneously, stability issues.[2] Clearly, the smaller the time step, the more stable the solution process, however, more time steps means more system evaluations. One would like to keep the time steps near the (Courant–Friedrichs–Levy) CFL limit, or slightly below it. The CFL condition dictates that numerical wave speed of $\Delta x/\Delta t$ must be at least as fast as the physical wave speed (c). The electromagnetic wave speed is $c = \frac{1}{\sqrt{\epsilon_o \mu_o}} \approx 2.99792458 \times 10^8 \pm 1.1\, m/s$ in vacuum while, in other media, it is $c = \frac{1}{\sqrt{\epsilon\mu}}$, where the corresponding dielectric constants are ϵ and μ. Thus, since $\Delta x/\Delta t \geq c$, this leads to the restriction, $\Delta t \leq \frac{\Delta x}{c}$. Because we are dealing with heterogeneous media in three dimensions, an ad hoc, and somewhat conservative, restriction is

$$\Delta t \leq \frac{\min(\Delta x_1, \Delta x_2, \Delta x_3)}{c_{\max}} \stackrel{\text{def}}{=} \Delta t^*, \tag{9.50}$$

where $c_{\max}$ is the fastest wave speed associated with the material components in the heterogeneous medium. Stability can, of course, be achieved by using an implicit scheme, which we employ. However, in many cases, this critical

[2] Typically, the number of iterations needed to solve the coupled system, if an iterative scheme is used, increases with the time step size and the value of ϕ.

condition (for explicit methods) still serves as an indicator of poor numerical behavior, *even for implicit schemes.* For the remainder of the work, we shall refer to the "CFL number" as

$$\mathcal{CFL} \stackrel{\text{def}}{=} \Delta t \left(\frac{c_{\max}}{\min(\Delta x_1, \Delta x_2 \Delta x_3)} \right) \stackrel{\text{def}}{=} \frac{\Delta t}{\Delta t^*}. \tag{9.51}$$

Although we will not employ explicit time-stepping schemes, we refer the reader to Kunz and Luebbers [53], and Taflove and Hagness [98] for overviews, in particular on the popular Yee-scheme (Yee [110]). We remark that a critical issue in the use of explicit schemes is stability, i.e., that errors at one time step do not grow in the next time step. We refer the reader to Taflove and Hagness [98] for a detailed discussion of this topic, and detailed analysis of the Yee method, which is subject to time step restrictions due to stability issues. We remark that, in the present work, since we shall iterate anyway due to the use of multifield staggering schemes, implicit methods are preferred for the applications of interest.

Remark: As the physics change, the field that is most sensitive (exhibits the largest amount of relative nondimensional change during the iterations) dictates the time step size. Because the internal system solvers within the staggering scheme are also iterative and use the previously converged solution as their starting value to solve the system of equations, a field that is relatively insensitive at given stage of the simulation will converge in very few internal iterations (perhaps even one).

9.4 Mesoscale Computations

9.4.1 Sample Size Selection

Generally, in order to determine $\boldsymbol{\epsilon}^*$ (for an anisotropic overall response), one specifies three uniform (spatially constant) linearly independent loadings (either $\boldsymbol{E}$ or $\boldsymbol{D}$) on the boundary of a sample. [3] Each independent loading yields three different averaged electric field components and hence provides three equations for the constitutive constants in $\boldsymbol{\epsilon}^*$. In order to determine the effective properties of the sample, one computes nine (actually 6 due to symmetry) constitutive constants ϵ^*_{ij} in the following relation between averages, that is,

$$\left\{ \begin{array}{c} \langle D_1 \rangle_\Omega \\ \langle D_2 \rangle_\Omega \\ \langle D_3 \rangle_\Omega \end{array} \right\} = \left[\begin{array}{ccc} \epsilon^*_{11} & \epsilon^*_{12} & \epsilon^*_{13} \\ \epsilon^*_{21} & \epsilon^*_{22} & \epsilon^*_{23} \\ \epsilon^*_{31} & \epsilon^*_{32} & \epsilon^*_{33} \end{array} \right] \left\{ \begin{array}{c} \langle E_1 \rangle_\Omega \\ \langle E_2 \rangle_\Omega \\ \langle E_3 \rangle_\Omega \end{array} \right\}. \tag{9.52}$$

If the effective response is assumed isotropic, then only one test loading (instead of usually three) is required, namely,

[3] In the time-transient case, the electrical and magnetic fields are coupled, and magnetic boundary data must also be supplied (either $\boldsymbol{H}$ or $\boldsymbol{B}$).

$$\epsilon^* \stackrel{\text{def}}{=} \sqrt{\frac{\langle \boldsymbol{D}\rangle_\Omega \cdot \langle \boldsymbol{D}\rangle_\Omega}{\langle \boldsymbol{E}\rangle_\Omega \cdot \langle \boldsymbol{E}\rangle_\Omega}}, \tag{9.53}$$

with a similar relation holding for the magnetic permeability,

$$\mu^* \stackrel{\text{def}}{=} \sqrt{\frac{\langle \boldsymbol{B}\rangle_\Omega \cdot \langle \boldsymbol{B}\rangle_\Omega}{\langle \boldsymbol{H}\rangle_\Omega \cdot \langle \boldsymbol{H}\rangle_\Omega}}. \tag{9.54}$$

Since we will be dealing with materials comprised of randomly dispersed particulate media, we shall assume that the materials have an overall isotropic response and that Equations 9.53 and 9.54 are adequate to describe the effective material. In order to select a suitable sample that is statistically representative (a RVE), we employ a "framing" method, whereby the uniform boundary (either $\boldsymbol{E}$ or $\boldsymbol{D}$) are applied to the boundary of a sample (Figure 9.5), and a subsample is used for the averaging process. This approach avoids introducing boundary layer effects into the volumetric averaging. An implementation of a "framing" approach is as follows:

- *Step (1):* Generate a sample with a certain number of particles in its interior to meet the volume fraction under investigation,
- *Step (2):* For the effective property calculation (averaging), select a subsample ("a sub-box," Figure 9.5) in the interior (to avoid boundary layer effects that arise from the imposition of uniform boundary conditions),
- *Step (3):* Repeat Steps (1) and (2) for different random realizations for a given sample size, and average the resulting effective properties to determine a mean value,
- *Step (4):* Repeat Steps (1)-(3) for a larger sample,
- *Step (5):* Continue the process (Steps (1)-(4)) until the effective property ceases to change to within an acceptable tolerance.

For a more in-depth discussion on size-effect issues, see the works of Zohdi et al. [120–128].

9.4.2 A Model Problem

As a model problem, we consider a heterogeneous material combination comprised of a group of particles in a binding matrix. We generated a group of N_p randomly dispersed spherical particles, of equal size, embedded in a cubical domain of dimensions, $D \times D \times D$. The particle size was determined by a particle/sample size ratio, which was defined via a subvolume size $V \stackrel{\text{def}}{=} \frac{D\times D\times D}{N_p}$. The nondimensional ratio between the radius (b) and the subvolume was denoted by $\mathcal{L} \stackrel{\text{def}}{=} \frac{b}{V^{\frac{1}{3}}}$. The volume fraction occupied by the particles consequently

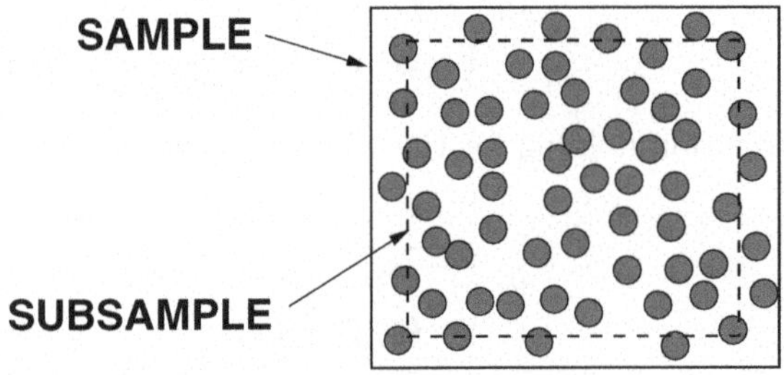

Fig. 9.5. With the framing method, a sample is probed with interior subsamples, within the larger sample, in order to avoid boundary effects that occur from imposing the uniform fields on the large-sample exterior (Zohdi [139]).

can be written as $v_p \stackrel{\text{def}}{=} \frac{4\pi\mathcal{L}^3}{3}$. Thus, the total volume occupied by the particles, denoted ζ, can be written as $\zeta = v_p N_p V$. We used $N_p = 100$ particles (Figure 9.6). This sample size was arrived at by successively enlarging sample until there were no significant changes in the overall system response for further enlargements. The classical random sequential addition algorithm was used to place nonoverlapping particles randomly into the domain of interest (RSA; Widom [108]). This algorithm was adequate for the volume fraction range of interest. However, if higher volume fractions are desired, more sophisticated algorithms, such as the well-known, equilibrium-based, Metropolis algorithm can be used. For even higher volume fractions (particle packing), a relatively recent class of efficient methods, based on simultaneous particle flow and growth, has been developed by Torquato and coworkers (Torquato [104], Kansaal et al. [49], and Donev et al. [19–21]).

9.4.3 System Parameters

The following parameters were used:

- A sample size of $D \times D \times D$, with $D = 0.00005\,m$, a subsample having a length-scale of $0.80 \times D$,
- An electric field on the boundary (linearly-growing), $\boldsymbol{E} = (10^9, 10^9, 10^9)\frac{t}{T}$, initial conditions, $\boldsymbol{E}(t=0) = (0, 0, 0)$,

- A magnetic field on the boundary (linearly-growing), $\boldsymbol{H} = (10^9, 10^9, 10^9)\sqrt{\frac{\mu_o}{\epsilon_o}\frac{t}{T}}$, initial conditions, $\boldsymbol{H} = (0, 0, 0)$,
- A time-stepping factor $\phi = 0.5$ (midpoint rule),
- A (nominal) length scale of the particles of $\zeta = 0.325$, with a $\pm$ 10 % variation of the resulting radius given afterwards,
- A temperature on the boundary, $\theta = 303.13^o \, Kelvin$, and initial conditions, $\theta = 303.13^o \, Kelvin$, along with a reference temperature of $\theta_o = 303.13 \, Kelvin$,
- A relative permittivity for the particles of $\epsilon_{2r} = 50$,
- A relative permittivity for the matrix of $\epsilon_{1r} = 2$,
- A relative permeability for the particles of $\mu_{2r} = 5$,
- A relative permeability for the matrix of $\mu_{1r} = 1$,
- An electrical conductivity for the particles of $\sigma_2 = 1$,
- An electrical conductivity for the matrix of $\sigma_1 = 1$,
- A density for the particles of $\rho_2 = 2000$,
- A density of the matrix of $\rho_1 = 1000$,
- A thermal conductivity for the particles of $I\!K_2 = 200$,
- A thermal conductivity for the matrix of $I\!K_1 = 100$,
- A heat capacity for the particles of $C_2 = 2000$,
- A heat capacity for the matrix of $C_1 = 1000$,
- An absorption for the particles of $a_2 = 1$,
- An absorption for the matrix of $a_1 = 1$,
- A (inner loop) solver tolerance, $tol = 10^{-6}$, and with the number of desired iterations per time step set to $K_d = 5$, along with a coupling/staggering tolerance of $C_{tol} = 10^{-6}$.

Remark: All of the Sigmoid function constants in Equations 9.1-9.5 were set to ± 1;

- *for the permittivity* :
(a)$\mathcal{K}_{E1} = 1$,
(b)$\alpha_{E1} = 1$,
(c)$\mathcal{K}_{E2} = 1$,
(d)$\alpha_{E2} = 1$,
- *for the permeability* :
(a)$\mathcal{K}_{H1} = 1$,
(b)$\alpha_{H1} = -1$,
(c)$\mathcal{K}_{H2} = 1$,
(d)$\alpha_{H2} = -1$,
- *and for electrical conductivity:*
(a)$\mathcal{K}_{S1} = 1$,
(b)$\alpha_{S1} = -1$,
(c)$\mathcal{K}_{S2} = 1$,
(d)$\alpha_{S2} = -1$.

Note that asymptotically, for the material behavior, as $\theta \to \infty$ and as $||\boldsymbol{E}|| \to \infty$,

$$\mathcal{E}(\theta - \theta_R, \boldsymbol{E} - \boldsymbol{E}_R) \to (1 + \mathcal{K}_{E1} + \mathcal{K}_{E2}) \tag{9.55}$$

and

$$\sigma(\theta_R, \boldsymbol{E}_R) \to 1, \tag{9.56}$$

while as $\theta \to \infty$ and as $||\boldsymbol{H}|| \to \infty$,

$$\mathcal{H}(\theta - \theta_R, \boldsymbol{H} - \boldsymbol{H}_R) \to 1. \tag{9.57}$$

The reference fields in the Sigmoid functions, $\boldsymbol{E}_R$ and $\boldsymbol{H}_R$, were both set to zero in the calculations.

9.4.4 Numerical Results

For the 100 particle sample, the meshes were repeatedly refined in the following sequential manner:

1. *Mesh # 1:* a 41 × 41 × 41 mesh[4] which has 413,526 *electromagnetic degrees of freedom* and 68,921 *thermodynamic degrees of freedom* for a total of 482,447 *degrees of freedom*,
2. *Mesh # 2:* a 61 × 61 × 61 mesh which has 1,361,886 *electromagnetic degrees of freedom* and 226,981 *thermodynamic degrees of freedom* for a total of 1,588,867 *degrees of freedom*,
3. *Mesh # 3:* a 81 × 81 × 81 mesh which has 3,188,646 *electromagnetic degrees of freedom* and 531,441 *thermodynamic degrees of freedom* for a total of 3,720,087 *degrees of freedom*,
4. *Mesh # 4:* a 101 × 101 × 101 mesh which has 6,181,806 *electromagnetic degrees of freedom* and 1,030,301 *thermodynamic degrees of freedom* for a total of 7,212,107 *degrees of freedom.*

Approximately, beyond the 61/81-level, there were no noticeable changes in the results. *All numerical results are shown in Figures 9.6-9.16.* At the length-scales of interest, it is questionable whether the ideas of a sharp material interface are justified. Accordingly, we simulated the system with and without Laplacian smoothing, whereby one smooths the material data by postprocessing the material data, node by node, to produce a smoother material representation, for example, for the electric permittivity, $\hat{\epsilon}$ (using the stencil in Figure 9.3),

[4] A 41 × 41 × 41 arises from having a cubical mesh with 20 nodes from the centerline line/plane of symmetry and one node in the middle.

$$\nabla_x^2 \epsilon = \mathbf{0} \Rightarrow \hat{\epsilon}_{i,j,k} = \frac{1}{6}\left(\epsilon_{i+1,j,k} + \epsilon_{i-1,j,k} + \epsilon_{i,j+1,k} + \epsilon_{i,j-1,k} + \epsilon_{i,j,k+1} + \epsilon_{i,j,k-1}\right). \tag{9.58}$$

The same was done for the permeability by enforcing $\nabla_x^2 \boldsymbol{\mu} = \mathbf{0}$, as well as other material data. The simulations were run with and without data smoothing, with the results being negligibly different for sufficiently fine meshes (Figure 9.6).

9.4.5 Observations

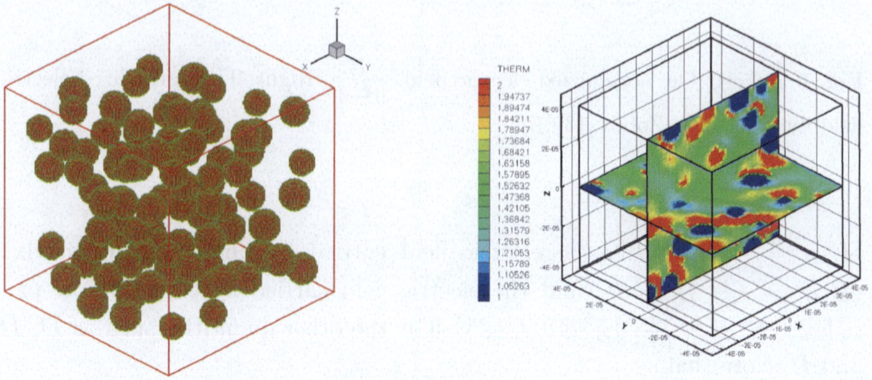

Fig. 9.6. Left: The morphology of the test sample's numerically-resolved microstructure, with a 101 × 101 × 101 mesh which has 6,181,806 *electromagnetic degrees of freedom*. Approximately beyond the 61/81 mesh-density level, there were no perceivable changes in the results. Right: the normalized temperature snapshot of $\frac{\theta}{\theta_o}$ during the simulation interval (Zohdi [139]).

The effective properties for both thermally-sensitive and thermally-insensitive are shown in Figures 9.6-9.16. Both the thermally-insensitive and thermally-sensitive cases produce effective responses which eventually fall within the bounds (using material parameters with all Sigmoid constants, the $\mathcal{K}$'s, set to zero). The thermally-sensitive case exhibits large overshoot due to the changing properties, however, it eventually converges (saturates) to a steady-state value.

The portion of the overall field carried by each phase can be an important quantity for a material design. The "load share" of the various fields can be post processed from a direct simulation. Figures 9.10-9.14 illustrate various overall field quantities, for example, the overall electric field:

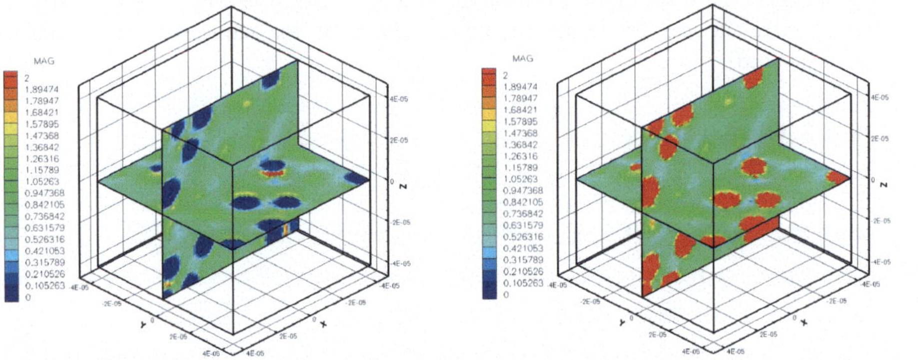

Fig. 9.7. Left: The normalized electric field, $\frac{\boldsymbol{E}}{||\boldsymbol{E}_o||}$. Right: The normalized electric field flux, $\frac{\boldsymbol{D}}{\epsilon_o||\boldsymbol{E}_o||}$ (Zohdi [139]).

$\langle \boldsymbol{E}\rangle_\Omega = \frac{1}{|\Omega|}\int_\Omega \boldsymbol{E}\, d\Omega$, the electric field carried by phase 1 (the matrix): $\langle \boldsymbol{E}\rangle_{\Omega_1} = \frac{1}{|\Omega_1|}\int_{\Omega_1} \boldsymbol{E}\, d\Omega$ and the electric field carried by phase 2 (the particles): $\langle \boldsymbol{E}\rangle_{\Omega_2} = \frac{1}{|\Omega_2|}\int_{\Omega_2} \boldsymbol{E}\, d\Omega$, as well as the other quantities such as $\boldsymbol{D}$, $\boldsymbol{H}$ and $\boldsymbol{B}$. Note that

$$\langle \boldsymbol{E}\rangle_\Omega = \frac{1}{|\Omega|}(\int_{\Omega_1} \boldsymbol{E}\, d\Omega + \int_{\Omega_2} \boldsymbol{E}\, d\Omega) = v_1\langle \boldsymbol{E}\rangle_{\Omega_1} + v_2\langle \boldsymbol{E}\rangle_{\Omega_2} \tag{9.59}$$

and

$$\langle \boldsymbol{D}\rangle_\Omega = \frac{1}{|\Omega|}(\int_{\Omega_1} \boldsymbol{D}\, d\Omega + \int_{\Omega_2} \boldsymbol{D}\, d\Omega) = v_1\langle \boldsymbol{D}\rangle_{\Omega_1} + v_2\langle \boldsymbol{D}\rangle_{\Omega_2}. \tag{9.60}$$

As illustrated in Figure 9.16, we note that the time steps were initially set to be the CFL limit $\mathcal{CFL} = 1$, but had to be refined below that level for the thermally-sensitive case. Unrefinements also took place, when allowable (when below the upper time step limit). The coupling tolerance was set to $C_{tol} = 10^{-6}$ in the preceding calculations. Only the thermally-sensitive problems required time step adaptivity (below the CFL limit) to control the coupling error. However, if the C_{tol} was made coarser, for example, to $C_{tol} = 10^{-5}$ and $C_{tol} = 10^{-4}$, etc., eventually the CFL-time step would have been adequate for the thermally-sensitive case, as for the thermally-insensitive case.

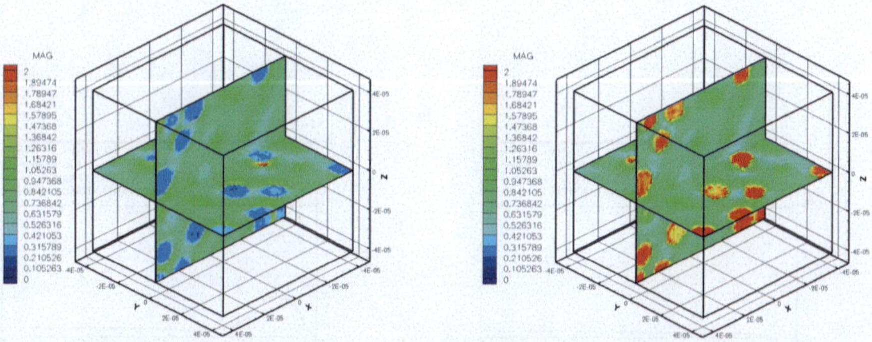

Fig. 9.8. Left: The normalized magnetic field, $\frac{\boldsymbol{H}}{||\boldsymbol{H}_o||}$. Right: The normalized magnetic field flux, $\frac{\boldsymbol{B}}{\mu_o||\boldsymbol{H}_o||}$ (Zohdi [139]).

Clearly, these results are dependent on the material parameters selected. It is important to stress that it is virtually impossible to determine a priori whether the initial time step is adequate to meet a tolerance and whether adaptivity is needed. *Obviously, we can use this scheme for any (trapezoidal) value of* $0 \leq \phi \leq 1$. Time step size adaptivity is important since the solution can dramatically change over the course of time, possibly requiring quite different time step sizes to control the iterative (staggering) error. However, to maintain the accuracy of the time-stepping scheme, one must respect an upper bound dictated by the discretization error, i.e., $\Delta t \leq \Delta t^{lim}$ (and the CFL condition). As stated before, classical solution methods require $\mathcal{O}(n^3)$ operations, whereas iterative schemes, such as the one presented, typically require order n^q, where $1 \leq q \leq 2$ and where n is the number of numerical unknowns. For details, see Axelsson [5]. Also, such solvers are highly advantageous since solutions to previous time steps can be used as the first guess to accelerate the solution procedure.

9.5 Stress- and Chemically-Induced Damage

The generation of heat and stress (due to Joule-heating and electromagnetically-induced body forces) can initiate forms of damage that are both mechanically-induced and chemically-induced. Accordingly, we now formulate extensions, involving stress- and chemically-induced degradation by enlarging the formulation to multifield systems to include simultaneous

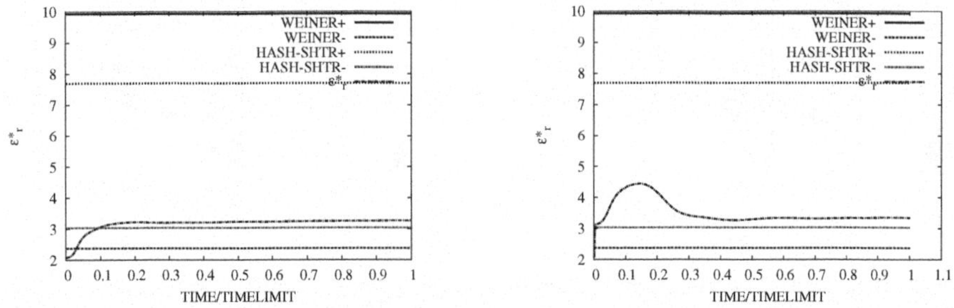

Fig. 9.9. The *isotropic* effective permittivity (ϵ^*). Left: Thermally-insensitive case. Right: Thermally-sensitive case. Note: The analytical bounds that are shown, are based on expressions having all constants set to zero in the Sigmoid functions (Zohdi [139]).

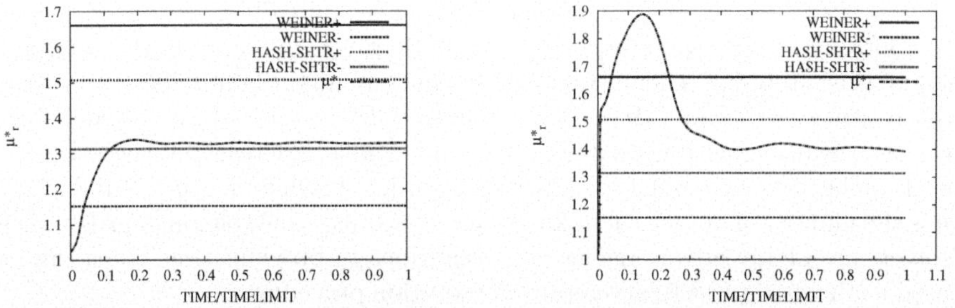

Fig. 9.10. The *isotropic* effective permeability (μ^*). Left: Thermally-insensitive case. Right: Thermally-sensitive case. Note: The analytical bounds that are shown are based on expressions having all constants set to zero in the Sigmoid functions (Zohdi [139]).

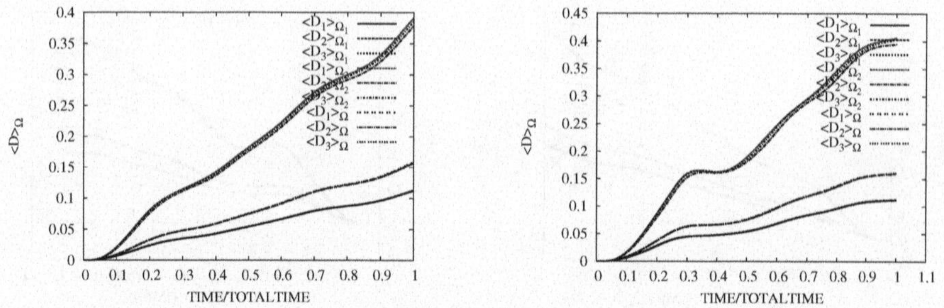

Fig. 9.11. The volume averaged electric flux: (1) carried by the particles $\langle \boldsymbol{D} \rangle_{\Omega_2}$ (top curves), (2) carried by the matrix $\langle \boldsymbol{D} \rangle_{\Omega_1}$ (bottom curves) and (3) overall $\langle \boldsymbol{D} \rangle_{\Omega}$ (middle curves). Left: Thermally-insensitive case. Right: Thermally-sensitive case. In all the cases, the overall average of the $\boldsymbol{D}$-components in each phase are virtually identical (equal in all three directions, as expected due to the the equiaxial boundary loading), i.e., the overall response ($\boldsymbol{\epsilon}^*$) is isotropic (Zohdi [139]).

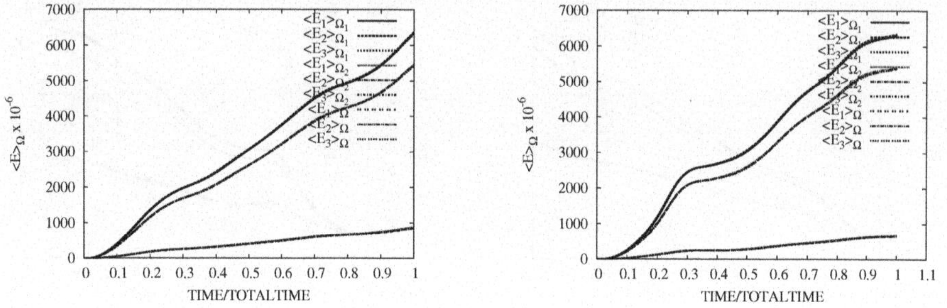

Fig. 9.12. The volume averaged electric field: (1) carried by the particles $\langle \boldsymbol{E} \rangle_{\Omega_2}$ (bottom curves), (2) carried by the matrix $\langle \boldsymbol{E} \rangle_{\Omega_1}$ (top curves) and (3) overall $\langle \boldsymbol{E} \rangle_{\Omega}$ (middle curves). Left: Thermally-insensitive case and Right: Thermally-sensitive case. In all the cases, the overall average of the $\boldsymbol{E}$-components in each phase are virtually identical (equal in all three directions, as expected due to the the equiaxial boundary loading), i.e., the overall response ($\boldsymbol{\epsilon}^*$) is isotropic (Zohdi [139]).

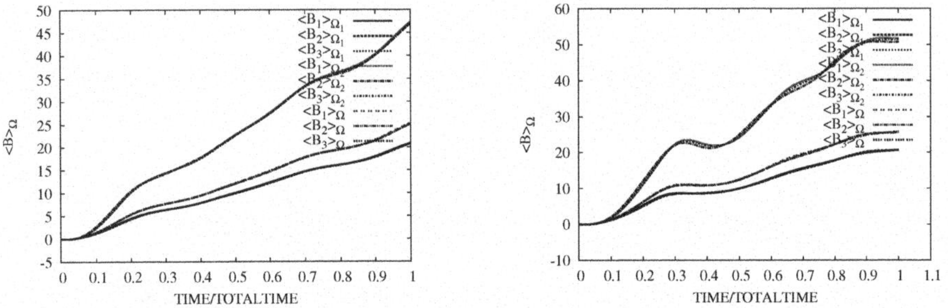

Fig. 9.13. The volume averaged magnetic flux: (1) carried by the particles $\langle \boldsymbol{B} \rangle_{\Omega_2}$ (top curves), (2) carried by the matrix $\langle \boldsymbol{B} \rangle_{\Omega_1}$ (bottom curves) and (3) overall $\langle \boldsymbol{B} \rangle_{\Omega}$ (middle curves). Left: Thermally-insensitive case. Right: Thermally-sensitive case. In all the cases, the overall average of the $\boldsymbol{D}$-components in each phase are virtually identical (equal in all three directions, as expected due to the the equiaxial boundary loading), i.e., the overall response ($\boldsymbol{\mu}^*$) is isotropic (Zohdi [139]).

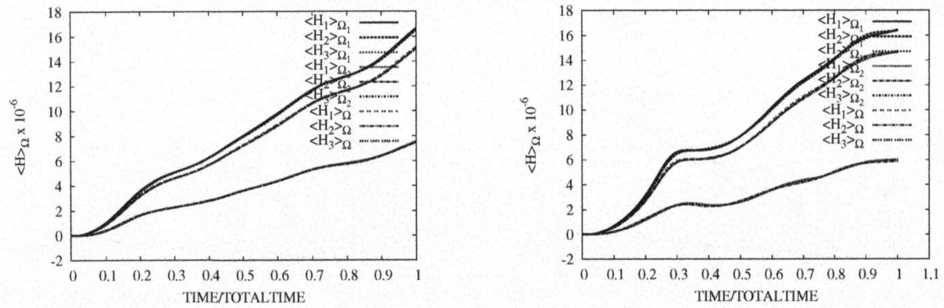

Fig. 9.14. The volume averaged magnetic field: (1) carried by the particles $\langle \boldsymbol{H} \rangle_{\Omega_2}$ (bottom curves), (2) carried by the matrix $\langle \boldsymbol{H} \rangle_{\Omega_1}$ (top curves) and (3) overall $\langle \boldsymbol{H} \rangle_{\Omega}$ (middle curves). Left: Thermally-insensitive case. Right: Thermally-sensitive case. In all the cases, the overall average of the $\boldsymbol{H}$-components in each phase are virtually identical (equal in all three directions, as expected due to the the equiaxial boundary loading), i.e., the overall response ($\boldsymbol{\mu}^*$) is isotropic (Zohdi [139]).

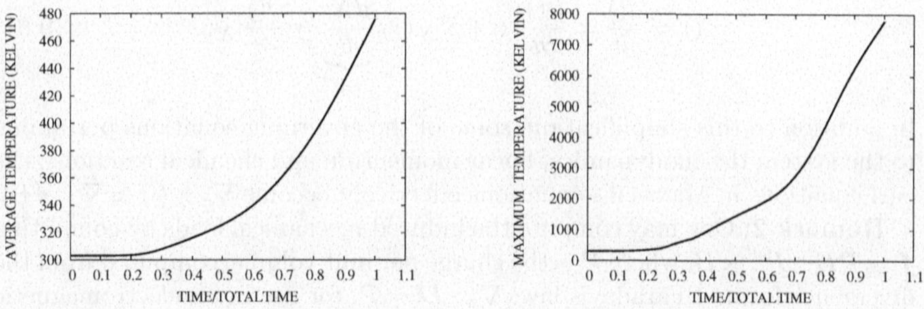

Fig. 9.15. Left: Average temperature ($\langle\theta\rangle_\Omega$). Right: The maximum temperature for the thermally-sensitive case (Zohdi [139]).

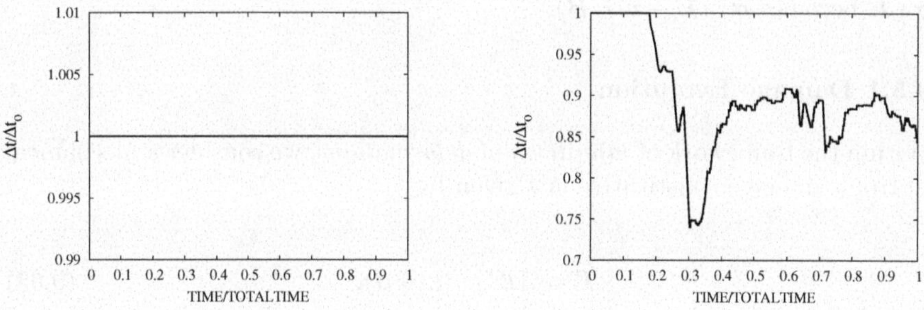

Fig. 9.16. Time step size relative to the $\mathcal{CFL}$-limit size (constant for the uncoupled problem). Left: Thermally-insensitive case. Right: Thermally-sensitive case. The time steps were initially set to be the CFL limit $\mathcal{CFL} \stackrel{\text{def}}{=} \Delta t \left(\frac{c_{\max}}{\min(\Delta x_1, \Delta x_2, \Delta x_3)} \right) = 1$, but had to be refined below that level (Zohdi [139]).

time-transient solution of: (1) Maxwell's equations, (2) The first law of thermodynamics, (3) The balance of linear momentum and (4) Reaction-diffusion laws.

Remark 1: We consider regimes where *infinitesimal deformations are appropriate.* Consistent with the infinitesimal deformation approximation, where $\boldsymbol{X}$ are referential coordinates, $\boldsymbol{x}$ are current coordinates and $\nabla_x(\cdot) \approx \nabla_X(\cdot)$ ($\frac{\partial(\cdot)}{\partial \boldsymbol{x}} \approx \frac{\partial(\cdot)}{\partial \boldsymbol{X}}$). The mechanical displacement is given by $\boldsymbol{u} = \boldsymbol{x} - \boldsymbol{X}$, and time derivatives given by

$$\dot{()} = \frac{d()}{dt} = \frac{\partial()}{\partial t}|_X + \nabla_X() \cdot \underbrace{\frac{d\boldsymbol{X}}{dt}}_{=0} = \frac{\partial()}{\partial t}|_X. \tag{9.61}$$

In addition to this simplification, some of the governing equations pertaining to the system thermodynamics, linear momentum and chemical reactions, the curl equations in Maxwell's equations effectively become $\nabla_x \times (\cdot) \approx \nabla_X \times (\cdot)$.

Remark 2: One may compute the induced mechanical loads by computing $\boldsymbol{f} = \mathcal{P}\boldsymbol{E} + \boldsymbol{J}^{tot} \times \boldsymbol{B}$, where $\mathcal{P}$ is the charge per unit volume, computed from the divergence form of Faraday's law, $\nabla_x \cdot \boldsymbol{D} = \mathcal{P}$, for use in an electromagnetic stress analysis, where

$$\nabla_x \cdot \boldsymbol{T} + \boldsymbol{f} = \rho \frac{d^2 \boldsymbol{u}}{dt^2}, \tag{9.62}$$

$\boldsymbol{T}$ being the Cauchy stress.

Remark 3: Note, for a deforming medium, in the Faraday's law, the term $\boldsymbol{\sigma} \cdot \boldsymbol{E}$ becomes $\boldsymbol{\sigma} \cdot (\boldsymbol{E} + \boldsymbol{v} \times \boldsymbol{B})$.

9.5.1 Damage Evolution

Within the framework of *infinitesimal deformations*, we consider a mechanical isotropic damage constitutive law given by

$$\boldsymbol{T} = \Lambda \mathbb{E}_0 : (\boldsymbol{\varepsilon} - \boldsymbol{\beta}), \tag{9.63}$$

where $\boldsymbol{T}$ is the Cauchy stress, governed by a balance of linear momentum [5]

$$\nabla_x \cdot \boldsymbol{T} + \boldsymbol{f} = \rho \ddot{\boldsymbol{u}} \tag{9.64}$$

which under infinitesimal deformation framework becomes

$$\nabla_X \cdot \boldsymbol{T} + \boldsymbol{f} = \rho_o \frac{\partial^2 \boldsymbol{u}}{\partial t^2}, \tag{9.65}$$

or, explicitly,

[5] The body force represents the electromagnetic coupling term, $\boldsymbol{f} = \mathcal{P}\boldsymbol{E} + \boldsymbol{J}^{tot} \times \boldsymbol{B}$.

$$\nabla_X \cdot (\Lambda \mathbb{E}_0 : (\boldsymbol{\varepsilon} - \boldsymbol{\beta})) + \boldsymbol{f} = \rho_o \frac{\partial^2 \boldsymbol{u}}{\partial t^2}, \tag{9.66}$$

with infinitesimal strains given by $\boldsymbol{\varepsilon} = \frac{1}{2}(\nabla_X \boldsymbol{u} + (\nabla_X \boldsymbol{u})^T)$ and thermal strains given by $\boldsymbol{\beta} = \boldsymbol{\varepsilon}_\theta \stackrel{\text{def}}{=} \boldsymbol{\gamma} \cdot (\theta - \theta_0)\mathbf{1}$. Here, the (isotropic) damaged elasticity tensor is $\mathbb{E} = \Lambda \mathbb{E}_0$, where $\mathbb{E}_0$ represents the "virgin" isotropic undamaged material, $0 \leq \Lambda \leq 1$ is the scalar continuity (isotropic damage) parameter (Kachanov [48]), $\Lambda(t = 0) = 1$ indicates the initial undamaged state and $\Lambda \to 0$ indicates a completely damaged state. The damage arising from mechanical and chemical sources is modeled as being governed by evolution overstress functions of the form $(0 < \Lambda \leq 1)$

$$\dot{\Lambda} = \left(\underbrace{a_1|\dot{c}|}_{\text{chemistry}} + \underbrace{a_2 H(m_1)\, m_1}_{\text{dilatation}} + \underbrace{a_3 H(m_2)\, m_2}_{\text{deviator}} \right) \Lambda, \tag{9.67}$$

where $m_1 \stackrel{\text{def}}{=} \frac{|tr\boldsymbol{T}/3|}{|tr\boldsymbol{T}_{crit}/3|} - 1$ and $m_2 \stackrel{\text{def}}{=} \frac{||\boldsymbol{T}'||}{||\boldsymbol{T}'_{crit}||} - 1$, the normalized concentration of the solute is c, given in molecules per unit volume, where $\boldsymbol{T}'_{crit} \stackrel{\text{def}}{=} k_1 \frac{\boldsymbol{T}'}{||\boldsymbol{T}'||}$, $\frac{tr\boldsymbol{T}_{crit}}{3} \stackrel{\text{def}}{=} k_2$, k_1 and k_2 being material constants, $\boldsymbol{T}' = \boldsymbol{T} - \frac{tr\boldsymbol{T}}{3}\mathbf{1}$. $H(\cdot)$ is the Heaviside function that is equal to zero when the argument is negative and is equal to unity otherwise. The material parameters (rate constants) a_1 through a_3, and k_1 and k_2 are spatially variable (different for each phase). For further details on these types of phenomenological (damage) formulations, the interested reader is referred to the seminal work of Kachanov [48]. Clearly, further evolution laws can be written for other material property changes, such as the thermal conductivity, dielectric properties, etc., although only changes in the mechanical property $\mathbb{E}$ are considered during the formulations to follow.[6]

9.5.2 Modification of the First Law of Thermodynamics

For a referential formulation of the first law of thermodynamics, we make the following infinitesimal strain approximation for the stored energy

$$\rho_o w = W \approx \tfrac{1}{2}(\boldsymbol{\varepsilon} - \boldsymbol{\beta}) : \mathbb{E} : (\boldsymbol{\varepsilon} - \boldsymbol{\beta}) + \rho_o C\theta, \tag{9.68}$$

which implies

$$\rho_o \dot{w} = \dot{W} = \tfrac{1}{2}(\boldsymbol{\varepsilon} - \boldsymbol{\beta}) : \dot{\mathbb{E}} : (\boldsymbol{\varepsilon} - \boldsymbol{\beta}) + (\dot{\boldsymbol{\varepsilon}} - \dot{\boldsymbol{\beta}}) : \mathbb{E} : (\boldsymbol{\varepsilon} - \boldsymbol{\beta}) + \rho_o C\dot{\theta}, \tag{9.69}$$

and thus the first law becomes

[6] In the case of material isotropy, $\mathbb{E} : \boldsymbol{\varepsilon} = \lambda tr\boldsymbol{\varepsilon} + 2\mu\boldsymbol{\varepsilon}$, where λ is the Lame parameter and μ is the shear modulus.

$$\rho_o C\dot{\theta} = \boldsymbol{T} : \dot{\boldsymbol{\beta}} - \tfrac{1}{2}(\boldsymbol{\varepsilon} - \boldsymbol{\beta}) : \dot{\mathbb{E}} : (\boldsymbol{\varepsilon} - \boldsymbol{\beta}) + \nabla_X \cdot (\mathbb{K} \cdot \nabla_X \theta) + \rho_o z \tag{9.70}$$

where Fourier's law, $\boldsymbol{q} = -\mathbb{K} \cdot \nabla_X \theta$, has been employed.

Remark: The chemical production of energy at a point is modeled as being related to the change in the rate of reaction, $\rho_o z = \eta|\dot{c}|$, which is discussed in detail in the next section, where η is a spatially variable material parameter. The parameter η is positive for exothermic reactions and negative for endothermic reactions.

9.5.3 Solid-State Diffusion-Reaction

The mass balance for a small diffusing species, denoted by the normalized concentration of the solute c (molecules per unit volume), in an arbitrary subvolume of material contained within Ω, denoted ω, consists of a storage term ($\dot{c}$), a reaction term ($\dot{s}$), and an inward normal flux term ($-\boldsymbol{q}_c \cdot \boldsymbol{n}$), leading to

$$\int_\omega (\dot{c} + \dot{s})\, d\omega = -\int_{\partial\omega} \boldsymbol{q}_c \cdot \boldsymbol{n}\, da. \tag{9.71}$$

It is a classical *stoichiometrically inexact* approximation to assume that the diffusing species reacts (is created or destroyed) in a manner such that the rate of production of the reactant (s) is directly proportional to the concentration of solute (c) itself and the rate of change of the solute (c) species,

$$\dot{s} = \tau_1 c + \tau_2 \dot{c}. \tag{9.72}$$

Here, $\tau_1 = r_1 e^{-\frac{Q_1}{R\theta}}$ and $\tau_2 = r_2 e^{-\frac{Q_2}{R\theta}}$, where r_1 and r_2 are rate constants, Q_1 and Q_2 are activation energies per mole of diffusive species, R is the universal gas constant and θ is the temperature in degrees Kelvin. Upon substitution of these relations into the conservation law for the diffusing species, and after using the divergence theorem, since the volume ω is arbitrary, one has a Fickian diffusion-reaction model in strong form, assuming a Fickian-type law, $\boldsymbol{q}_c = -\mathbb{D} \cdot \nabla_X c$

$$\dot{c} = \nabla_X \cdot (\mathbb{D} \cdot \nabla_X c) - \tau_1 c - \tau_2 \dot{c} \Rightarrow \dot{c}(1 + \tau_2) = \nabla_X \cdot (\mathbb{D} \cdot \nabla_X c) - \tau_1 c, \tag{9.73}$$

where $\mathbb{D}$ is the diffusivity and, as before, $\dot{()} = \frac{\partial()}{\partial t}|_X$. When $r_1 > 0$, the diffusing species is destroyed as it reacts, while $r_1 < 0$ indicates that the diffusing species is created as it reacts, i.e., an autocatalytic or "chain" reaction occurs. Also, depending on the sign of r_2, effectively the process will have an accelerated or decelerated diffusivity as well as accelerated or decelerated reactivity. In Equation 9.73, the familiar Arrhenius form $\mathbb{D} = \mathbb{D}_0 e^{-\frac{U}{R\theta}}$ is implicitly used, where $\mathbb{D}_0$ is the diffusivity tensor at a reference temperature, and U is the activation energy per mole of diffusive species.

Remark 1: It is sometimes observed that, in regions of relatively high positive triaxial stress, the diffusion is accelerated, while in regions of high negative triaxial stress, diffusion is decelerated. Diffusion models with explicit pressure dependency will not be considered, however, we remark that a particularly simple constitutive model to incorporate stress-dependency phenomena is given by a pseudo-Fickian/Arrhenius law, $\boldsymbol{q}_c = -\mathbb{D}_0 e^{\frac{-U(\boldsymbol{T})}{R\theta}} \cdot \nabla c$, motivated by thermodynamical arguments found in the classical works of Flynn [30] or Crank [11]. [7]

Remark 2: It is important to note that instabilities can be induced by diffusion, i.e., a coupled mechanochemical system can be stable when no diffusion is present and unstable in the presence of diffusion. An in-depth mathematical analysis of such effects has been conducted by Markenscoff [65-67].

9.5.4 Discretization of the Mechanical and Concentration Fields

Mechanical field: For the mechanical field (infinitesimal deformation formulation), we write ($\boldsymbol{v}$ is the velocity)

$$\frac{d\boldsymbol{v}}{dt} = \frac{\partial \boldsymbol{v}}{\partial t} = \frac{1}{\rho_o}(\nabla_X \cdot \boldsymbol{T} + \boldsymbol{f}) \stackrel{\text{def}}{=} \boldsymbol{L}. \tag{9.74}$$

We discretize for time$=t + \phi \Delta t$, and using a trapezoidal "$\phi - scheme$" ($0 \leq \phi \leq 1$),

$$\frac{\boldsymbol{v}(t+\Delta t) - \boldsymbol{v}(t)}{\Delta t} \approx \boldsymbol{L}(t + \phi \Delta t) \approx \phi \boldsymbol{L}(t + \Delta t) + (1-\phi)\boldsymbol{L}(t). \tag{9.75}$$

Rearranging yields

$$\boldsymbol{v}(t+\Delta t) \approx \boldsymbol{v}(t) + \Delta t \left(\phi \boldsymbol{L}(t+\Delta t) + (1-\phi)\boldsymbol{L}(t)\right), \tag{9.76}$$

where the previously introduced spatial discretization is applied to the divergence terms ($\nabla_x \cdot \boldsymbol{T}$) in $\boldsymbol{L}$. Since this is a second order system, the procedure is then repeated to determine the displacement field $\boldsymbol{u}$.

Remark: As in our treatment of Maxwell's equations, finite difference stencils are used (with additional mixed derivatives), and the details were given earlier.

Chemical field: For the concentration field, we write

$$\frac{\partial c}{\partial t} = \frac{1}{1+\tau_2}(-\nabla_X \cdot \boldsymbol{q}_c - \tau_2 c) \stackrel{\text{def}}{=} Z. \tag{9.77}$$

---

[7] An additive split for stress dependency of the form $U(\boldsymbol{T}) = U_0 + \tilde{U}(p)$, where U_0 is a stress-independent reference activation energy and $p = \frac{tr\boldsymbol{T}}{3}$ is the pressure, has been given in Zohdi [116–119] for certain applications.

We discretize around the time=$t + \phi \Delta t$, yielding

$$c(t + \Delta t) \approx c(t) + \Delta t \left(\phi Z(t + \Delta t) + (1 - \phi) Z(t) \right), \tag{9.78}$$

where the previously introduced spatial discretization is applied to the divergence terms in $\boldsymbol{q}_c$.

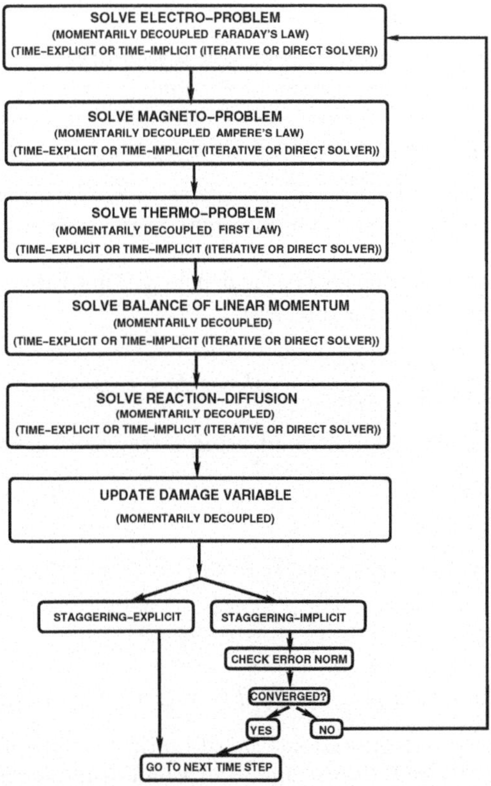

Fig. 9.17. Types of coupled staggering solution for the thermo-electro-magneto-mechano-chemo system (Zohdi [139]).

9.5.5 Extended Numerics: Staggering for Electro-Magneto-Thermo-Mechano-Chemo Systems

Extending the previous electro-magneto-thermo staggering scheme to include mechanical and chemical effects, one computes the $\boldsymbol{E}$-field with $\boldsymbol{H}$, θ, $\boldsymbol{u}$ and

c fields fixed, then computes $\boldsymbol{H}$-field with $\boldsymbol{E}$, θ, $\boldsymbol{u}$ and c fields fixed, etc., and iterates at time interval $L+1$, $K=1,2,...$ for (written directly in iterative implicit form)

$$\boldsymbol{E}^{L+1,K} = \mathcal{F}(\underline{\boldsymbol{E}^{L+1,K-1}}, \boldsymbol{H}^{L+1,K-1}, \theta^{L+1,K-1}, \boldsymbol{u}^{L+1,K-1}, c^{L+1,K-1}), \tag{9.79}$$

$$\boldsymbol{H}^{L+1,K} = \mathcal{G}(\boldsymbol{E}^{L+1,K}, \underline{\boldsymbol{H}^{L+1,K-1}}, \theta^{L+1,K-1}, \boldsymbol{u}^{L+1,K-1}, c^{L+1,K-1}), \tag{9.80}$$

$$\theta^{L+1,K} = \mathcal{Y}(\boldsymbol{E}^{L+1,K}, \boldsymbol{H}^{L+1,K}, \underline{\theta^{L+1,K-1}}, \boldsymbol{u}^{L+1,K-1}, c^{L+1,K-1}), \tag{9.81}$$

$$\boldsymbol{u}^{L+1,K} = \mathcal{L}(\boldsymbol{E}^{L+1,K}, \boldsymbol{H}^{L+1,K}, \theta^{L+1,K}, \underline{\boldsymbol{u}^{L+1,K-1}}, c^{L+1,K-1}), \tag{9.82}$$

and

$$c^{L+1,K} = \mathcal{C}(\boldsymbol{E}^{L+1,K}, \boldsymbol{H}^{L+1,K}, \theta^{L+1,K}, \boldsymbol{u}^{L+1,K}, \underline{c^{L+1,K-1}}), \tag{9.83}$$

where the only underlined variable is "active" at that stage of the process. One then compute error measures: $\varpi^{*,K} \stackrel{\text{def}}{=} max(\varpi_E^K, \varpi_H^K, \varpi_\theta^K, \varpi_u^K, \varpi_c^K)$, $i=1,...,nodes$ in the system, where, in addition to the previously introduced *nondimensional* errors for the electric, magnetic and thermal fields, we define, for the displacement,

$$\varpi_u^K \stackrel{\text{def}}{=} \frac{||\boldsymbol{u}^{L+1,K} - \boldsymbol{u}^{L+1,K-1}||}{||\boldsymbol{u}^{L+1,K} - \boldsymbol{u}^L||}, \tag{9.84}$$

and for the concentration,

$$\varpi_c^K \stackrel{\text{def}}{=} \frac{||c^{L+1,K} - c^{L+1,K-1}||}{||c^{L+1,K} - c^L||}. \tag{9.85}$$

Thereafter, we select the maximum error for adaptivity

$$\varpi^{*,K} \stackrel{\text{def}}{=} max(\varpi_E^K, \varpi_H^K, \varpi_\theta^K, \varpi_u^K, \varpi_c^K), \tag{9.86}$$

and proceed as introduced earlier for the electro-magneto-thermo scheme, with the modified flowchart shown in Figure 9.17.

Remark 1: An electro-magneto-thermo-mechano-chemo numerical example is provided shortly.

Remark 2: Because the temperatures can become quite high (as seen in Figures 9.6-9.16), especially without a cooling mechanism, one may wish to include phase transformations by considering four cases:

- *No melting:* If $\theta(t) < \theta_m$ and $\theta(t+\Delta t) < \theta_m$, then $C(\theta) = C_o$, where C_o is the solid heat capacity.

- *Melting:* If $\theta(t) < \theta_m$ and $\theta(t+\Delta t) \geq \theta_m$, then $C(\theta) = C_o + \frac{\Delta H_m^{S\to L}}{\Delta \theta_m}$, where $\Delta H_m^{S\to L}$ is the latent heat of melting. This has the effect of enforcing a constant temperature (or absorbing the latent heat), where $\Delta\theta_m$ is small and can be thought of as a "bandwidth" for melting.
- *Melted:* If $\theta(t) \geq \theta_m$ and $\theta(t+\Delta t) \geq \theta_m$, then $C(\theta) = C_m$, where C_m is the heat capacity of the melted material.
- *Solidification:* If $\theta(t) \geq \theta_m$ and $\theta(t+\Delta t) < \theta_m$, then $C(\theta) = C_m + \frac{\Delta H_m^{L\to S}}{\Delta \theta_m}$, where $\Delta H_m^{L\to S}$ is the latent heat of solidification.

This is relatively straightforward to include within the staggering framework.

9.6 Summary

The overall goal was to deliver solutions where the iterative staggering (incomplete coupling) error is controlled and the temporal discretization accuracy dictates the upper limits on the time step size (Δt^{lim}). Generally speaking, the staggering error, which is a function of the time step size, is time-dependent and can become stronger, weaker, or possibly oscillatory, is extremely difficult to ascertain a priori as a function of the time step size. Therefore, to circumvent this problem, the presented adaptive staggering strategy was developed to provide accurate solutions by iteratively adjusting the time steps. Specifically, a sufficient condition for the convergence of the presented fixed-point scheme was that the spectral radius (contraction constant of the coupled operator), which depends on the time step size, must be less than unity. This observation was used to adaptively control the time step sizes, while simultaneously controlling the coupled operator's spectral radius, in order to deliver solutions below an error tolerance within a prespecified number of desired iterations. This recursive staggering error control can allow for substantial reduction of computational effort by the adaptive use of large time steps, when possible. Furthermore, such a recursive process has a reduced sensitivity (relative to an explicit staggering approach) to the order in which the individual equations are solved since it is self-correcting.

Remark: Some examples of the application of computational electromagnetics simulations to biological problems, specifically red blood cell hemoglobinapathies, are provided later for the interested reader.

9.7 An Electro-Magneto-Thermo-Mechano-Chemo Numerical Example

As an example, in addition to the previous electromagnetic, the following parameters were used (see the equations in the main body of the work for variable definitions)[8]:

- The displacement boundary condition: $\boldsymbol{u}|_{\partial\Omega} = \boldsymbol{0}$,
- The (interior) displacement initial condition: $\boldsymbol{u}(x_1, x_2, x_3) = \boldsymbol{0}$,
- The chemical concentration boundary condition: $c|_{\partial\Omega} = 1000$,
- The (interior) chemical concentration initial condition: $c(x_1, x_2, x_3) = 0$,
- The matrix material shear modulus $\mu_m^{sh} = 50$ GPa,
- The particle material shear modulus $\mu_p^{sh} = 100$ GPa,
- The matrix material Lame parameter $\lambda_m = 100$ GPa,
- The particle material Lame parameter $\lambda_p = 200$ GPa,
- The matrix material thermal expansion coefficient $\beta_m = 10^{-6}$,
- The particle material thermal expansion coefficient $\beta_p = 5 \times 10^{-6}$,
- The matrix material chemical heat generation parameter (for the first law of thermodynamics) $\eta_m = 1000$; recall $\rho_o z = \eta|\dot{c}|$,
- The particle material chemical heat generation parameter $\eta_p = 5000$,
- The matrix material chemical damage rate parameter $a_{1m} = -10$,
- The particle material chemical damage rate parameter $a_{1p} = -50$,
- The matrix material pressure damage rate parameter $a_{2m} = -10$,
- The particle material pressure damage rate parameter $a_{2p} = -50$,
- The matrix material deviatoric damage rate parameter $a_{3m} = -10$,
- The particle material deviatoric damage rate parameter $a_{3p} = -50$,
- The matrix material critical pressure parameter $k_{1m} = 10$ MPa,
- The particle material critical pressure parameter $k_{1p} = 50$ MPa,
- The matrix material critical deviatoric norm parameter $k_{2m} = 10$ MPa,
- The particle material critical deviatoric norm $k_{2p} = 50$ MPa,
- The matrix material first reaction rate parameter $r_{1m} = 10^{-6}$,
- The particle material reaction rate parameter $r_{1p} = 5 \times 10^{-6}$,
- The matrix material reaction rate parameter $r_{2m} = 10^{-6}$,
- The particle material reaction rate parameter $r_{2p} = 5 \times 10^{-6}$,
- The matrix material baseline diffusivity $D_m = 1$,
- The particle material baseline diffusivity $D_p = 5$,
- The matrix material activation energy parameters for the reaction rate (r_{1m}): $Q_{11m} = 100$, $Q_{12m} = 10^{-6}$ for the following $\tau_1 = r_1 e^{\frac{Q_{11} - Q_{12}P}{R\theta}}$,
- The particle material activation energy parameters for the reaction rate (r_{1p}): $Q_{11p} = 500$, $Q_{12p} = 5 \times 10^{-6}$,

[8] The thermally-sensitive case was considered here. Also, the matrix material was signified with a subscript "m" and the particle material with a subscript "p" for clarity.

- The matrix material activation energy parameters for the reaction rate (r_{2m}): $Q_{21m} = 100$, $Q_{22m} = 10^{-6}$ for the following $\tau_2 = r_2 e^{\frac{Q_{21}-Q_{22P}}{R\theta}}$,
- The particle material activation energy parameters for the reaction rate (r_{2p}): $Q_{21p} = 500$, $Q_{22p} = 5 \times 10^{-6}$,
- The matrix material activation energy parameters for the diffusion (D_{om}): $U_{11m} = 100$, $U_{12m} = 10^{-6}$ for the following $\mathbb{D} = \mathbb{D}_o e^{\frac{U_{21}-U_{22P}}{R\theta}}$,
- The particle material activation energy parameters for the reaction rate (D_{op}): $U_{21p} = 500$, $U_{22p} = 5 \times 10^{-6}$.

The results are shown in Figures 9.18-9.21.

9.8 Thermoelectricity and More Coupling

Some brief remarks on thermoelectricity are in order. Thermoelectricity deals with the conversion of thermal energy into electrical energy, and vice versa. Unlike Joule-heating, the effects can be reversible. Essentially, one can generate current from a temperature difference. There are three main effects in this area:

- The *Seebeck effect* is characterized by an electrical field, $\boldsymbol{E}_s$, being produced by a temperature gradient

$$\boldsymbol{E}_s = \boldsymbol{S} \cdot \nabla_x \theta = \underbrace{S \nabla_x \theta}_{\text{isotropic}} . \tag{9.87}$$

where **S** is the Seebeck coefficient.
Maxwell's equations become modified to read; Faraday's law (no explicit change)

$$\nabla_x \times \boldsymbol{E} = -\left(\frac{\partial(\boldsymbol{\mu} \cdot \boldsymbol{H})}{\partial t} + \boldsymbol{M}^{ext} + \hat{\boldsymbol{\sigma}} \cdot \boldsymbol{H}\right) \tag{9.88}$$

and Ampere's law

$$\nabla_x \times \boldsymbol{H} = \frac{\partial(\boldsymbol{\epsilon} \cdot \boldsymbol{E})}{\partial t} + \boldsymbol{J}^{ext} + \boldsymbol{\sigma} \cdot (\boldsymbol{E} + \boldsymbol{E}_s). \tag{9.89}$$

- The *Thomson (Lord Kelvin) effect* describes a thermal source/sink in the first law that is proportional to the current and temperature gradient:

$$\rho \dot{w} = \boldsymbol{T} : \nabla_x \boldsymbol{v} - \nabla_x \cdot \boldsymbol{q} + a\boldsymbol{J} \cdot \boldsymbol{E} - \eta \boldsymbol{J} \cdot \nabla_x \theta, \tag{9.90}$$

where η is the Thomson coefficient, which is related to the Seebeck coefficient via

$$\eta = \theta \frac{\partial S}{\partial \theta}. \tag{9.91}$$

The Thomson coefficient can be positive or negative.

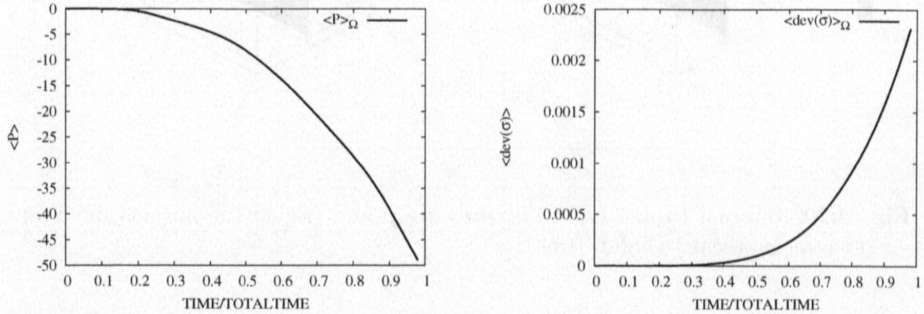

Fig. 9.18. Left: The volume averaged pressure $\langle \boldsymbol{p} \rangle_\Omega$. Right: The volume averaged normed deviator $\langle ||\boldsymbol{\sigma}'|| \rangle_\Omega$ (Zohdi [139]).

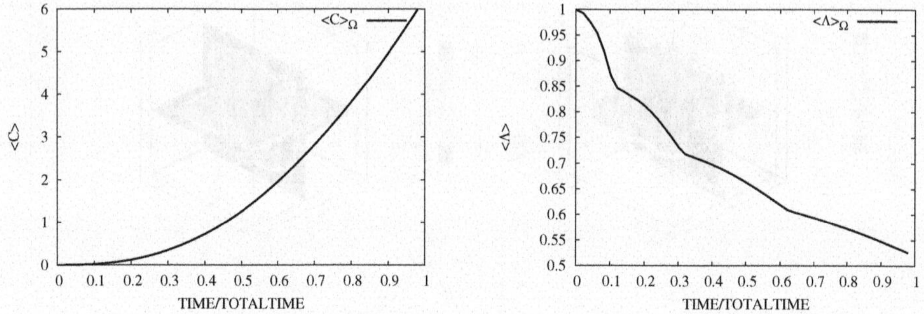

Fig. 9.19. Left: The volume averaged concentration $\langle c \rangle_\Omega$. Right: The volume averaged damage indicator $\langle \Lambda \rangle_\Omega$ (Zohdi [139]).

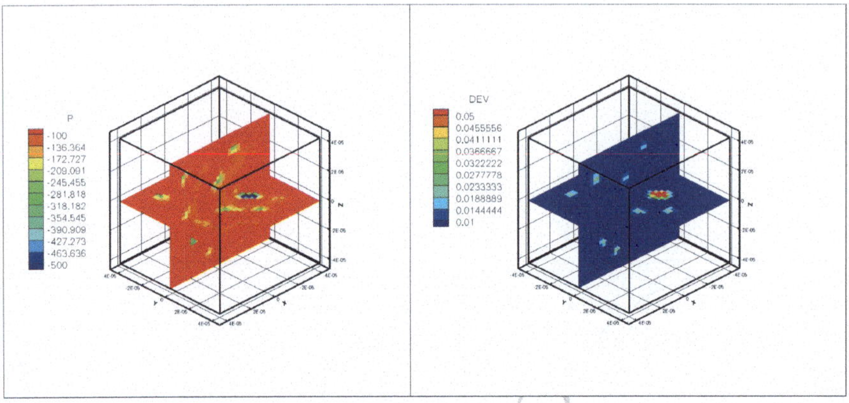

Fig. 9.20. Internal Probe: Left: The pressure $\boldsymbol{p}$ and (right) the normed deviator $||\boldsymbol{\sigma}'||$ (in megapascals) (Zohdi [139]).

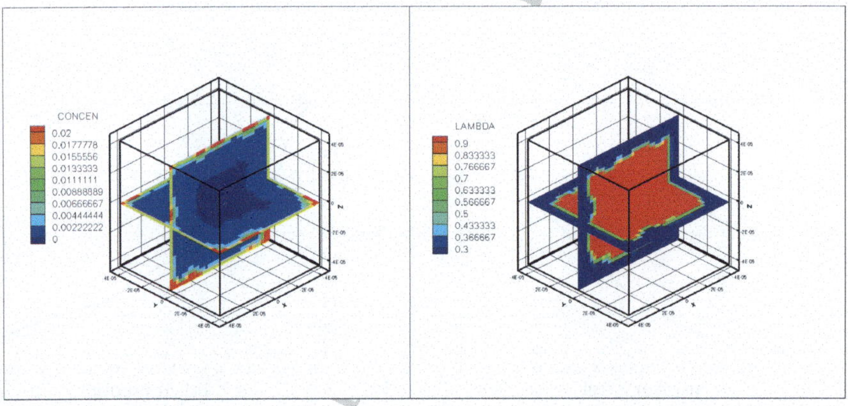

Fig. 9.21. Internal Probe: Left: The chemical concentration $\boldsymbol{c}$ and (right) the damage indicator Λ (Zohdi [139]).

- The *Peltier effect* states that the heat absorbed or generated (dependent on the direction of current flow) is proportional to the current in certain materials. The degree to which this effect is present in a material is given by the Peltier coefficient

$$\Pi = S\theta. \tag{9.92}$$

For both power generation and cooling, the thermoelectric material needs a high electrical conductivity, a high Seebeck coefficient and a low thermal conductivity. Accordingly, often, so-called "Figures of Merit" are defined via

$$\mathcal{Z} = \frac{S\sigma}{K}. \tag{9.93}$$

9.8.1 Temporal Discretization of the Coupled System

The discretization of this system follows the same lines as before, namely,

$$\frac{\partial(\boldsymbol{\mu} \cdot \boldsymbol{H})}{\partial t} = -\nabla_x \times \boldsymbol{E} - \boldsymbol{M}^{ext} - \hat{\boldsymbol{\sigma}} \cdot \boldsymbol{H} \stackrel{\text{def}}{=} \boldsymbol{F} \tag{9.94}$$

and

$$\frac{\partial(\boldsymbol{\epsilon} \cdot \boldsymbol{E})}{\partial t} = \nabla_x \times \boldsymbol{H} - \boldsymbol{J}^{ext} - \boldsymbol{\sigma} \cdot \boldsymbol{E} - \boldsymbol{\sigma} \cdot \boldsymbol{E}_s. \stackrel{\text{def}}{=} \boldsymbol{G}, \tag{9.95}$$

where $\boldsymbol{E}_s = S\nabla_x\theta$. We discretize for time$=t + \phi\Delta t$, and using a trapezoidal "$\phi - scheme$" ($0 \leq \phi \leq 1$),

$$\frac{(\boldsymbol{\mu} \cdot \boldsymbol{H})(t + \Delta t) - (\boldsymbol{\mu} \cdot \boldsymbol{H})(t)}{\Delta t} \approx \boldsymbol{F}(t+\phi\Delta t) \approx \phi\boldsymbol{F}(t+\Delta t)+(1-\phi)\boldsymbol{F}(t) \tag{9.96}$$

and

$$\frac{(\boldsymbol{\epsilon} \cdot \boldsymbol{E})(t + \Delta t) - (\boldsymbol{\epsilon} \cdot \boldsymbol{E})(t)}{\Delta t} \approx \boldsymbol{G}(t+\phi\Delta t) \approx \phi\boldsymbol{G}(t+\Delta t)+(1-\phi)\boldsymbol{G}(t). \tag{9.97}$$

Rearranging yields

$$\boldsymbol{H}(t + \Delta t) \approx \boldsymbol{\mu}^{-1}(t + \Delta t) \cdot ((\boldsymbol{\mu} \cdot \boldsymbol{H})(t) + \Delta t\,(\phi\boldsymbol{F}(t + \Delta t) + (1 - \phi)\boldsymbol{F}(t))) \tag{9.98}$$

and

$$\boldsymbol{E}(t + \Delta t) \approx \boldsymbol{\epsilon}^{-1}(t + \Delta t) \cdot ((\boldsymbol{\epsilon} \cdot \boldsymbol{E})(t) + \Delta t\,(\phi\boldsymbol{G}(t + \Delta t) + (1 - \phi)\boldsymbol{G}(t)))\,, \tag{9.99}$$

where the previously introduced spatial discretization is applied to the terms in $\boldsymbol{F}$ and $\boldsymbol{G}$ ($\nabla_x \times \boldsymbol{H}$ and $\nabla_x \times \boldsymbol{E}$).

For the thermal field, we write

$$\frac{\partial \theta}{\partial t} = \frac{1}{\rho C} \left(-\nabla_x \cdot \boldsymbol{q} + \rho z \right) \stackrel{\text{def}}{=} Y, \tag{9.100}$$

where $\rho z = a\boldsymbol{J} \cdot \boldsymbol{E} - \eta \boldsymbol{J} \cdot \nabla_x \theta +$chemical effects etc. and $\eta = \theta \frac{\partial S}{\partial \theta}$. We discretize around the time$=t + \phi \Delta t$, yielding

$$\theta(t + \Delta t) \approx \theta(t) + \Delta t \left(\phi Y(t + \Delta t) + (1 - \phi) Y(t) \right), \tag{9.101}$$

where the previously introduced spatial discretization is applied to the terms in Y. Then, the computations proceed just as in the previous chapter.

Remark: In modern day computing, there are a variety of techniques which seek to harness the homogenized properties of a material in order to help solve a boundary value problem on a nontrivial domain. These techniques can be collectively/generically described by:

1. Solve a "homogenized" (using homogenized, smooth, material coefficients $\boldsymbol{\mu}^o = \mu^*$ and $\boldsymbol{\epsilon}^o = \boldsymbol{\epsilon}^*$) boundary value problem: $\mathcal{A}^o(\boldsymbol{\epsilon}^o, \boldsymbol{\mu}^o, \boldsymbol{E}^o, \boldsymbol{D}^o, \boldsymbol{H}^o, \boldsymbol{B}^o) = \mathcal{F}$ to generate a homogenized solution $\boldsymbol{u}^o \stackrel{\text{def}}{=} (\boldsymbol{E}^o, \boldsymbol{D}^o, \boldsymbol{H}^o, \boldsymbol{B}^o)$.
2. Plug the homogenized solution $\boldsymbol{u}^o \stackrel{\text{def}}{=} (\boldsymbol{E}^o, \boldsymbol{D}^o, \boldsymbol{H}^o, \boldsymbol{B}^o)$ into the true governing equations: $\mathcal{A}(\boldsymbol{\epsilon}, \boldsymbol{\mu}, \boldsymbol{E}^o, \boldsymbol{D}^o, \boldsymbol{H}^o, \boldsymbol{B}^o) - \mathcal{F} = \mathcal{R}$.
3. Check $||\mathcal{R}|| = \sqrt{\mathcal{R} \cdot \mathcal{R}}$ and compare against a modeling error tolerance TOL.
4. For all locations where $||\mathcal{R}|| \geq TOL$, a local boundary value problem is solved, with boundary data for of the local domain.

This approach allows one to solve the computationally intensive "real" boundary value problem only where needed, and retains the computationally inexpensive homogenized solution in all other locations. This process has been studied *extensively by the author,* but is outside the scope of the present work (see Zohdi [114, 115] and Oden et al. [75]).

9.9 Electrically-Aided Sintering

An area where the previous analysis can play a very significant role is in electrically-aided sintering, that is, a subfield of powder processing. Powder processing refers to compacting powders in dies and sintering the particles together by heating to near melting. The process is applied to a variety of powdered metals, ceramics, glasses, graphites and diamond. The powders can be produced in a variety of ways, for example:

- atomization of liquid streams into droplets by breaking jets of metal,
- reduction of metal oxides and
- comminution/pulverizing of bulk material.

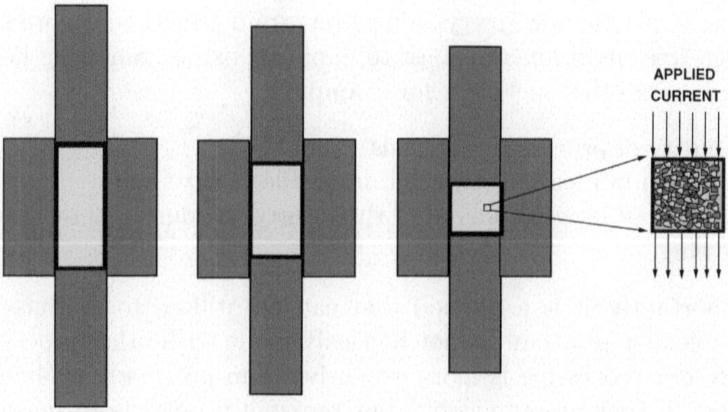

Fig. 9.22. A sequential compaction of powdered material with simultaneous applied current (Zohdi[140]).

The particles are usually passed through a series of sieves to separate particles of various sizes. Typically, the particles are poured with lubricants and binder mixtures into a mold to make for easy flow. Various blends of particle mixtures are frequently employed. The "as poured" mixture is referred to as a "green compact" before sintering. The three stages of material states are

- loose powder,
- green compact and
- sintered (bonded) state.

The "green" density depends on

- compaction pressure,
- powder composition and
- hardness of powder.

Compaction pressures usually range from 50-1000 Mpa, which is achieved by hydraulic presses and isostatic pressing. Afterwards, processes such as metal injection (into the interstitial pores), rolling, extrusion, etc., are applied. Sintering specifically refers to processing the compacted (green) material, which is brittle, and heating it to $70 - 90\%$ of melting. The methods by which this heat is delivered is wide ranging, for example, by placing it in a furnace with three chambers: (1) a burn off chamber to the vaporize lubricants, (2) a high temperature chamber to sinter and (3) a cooling chamber to ramp down the temperature. The binding occurs by small-scale mechanisms involving diffusion, plastic flow, recrystallization, grain growth and pore shrinkage. An oxygen-free environment is best to minimize oxides. Sintering has distinct advantages over other methods, for example:

- high purity of processed materials,
- few steps in fabrication (thus retaining the purity) and
- production of near net-shape of the desired product without much post-processing.

Most importantly, it is a method that can be utilized to produce products with complex shapes that cannot be easily made with other processes. Generally, powder processing is more expensive than processes involving molten equivalents, if they are applicable, but research is ongoing to make the process cheaper to expand the methods beyond niche markets. Electrically-aided sintering is one promising enhancement to the overall process. This involves placing loose powders in a graphite mold, which is heated by an electrical current and compacted, simultaneously (Figure 9.22). The quality of the properties of the bulk materials scale as a function of the pore volume fraction and the strength of the bonds. The key control quantity is the heat generated from an electrical field (Joule-heating). The interconversions of various forms of energy (electromagnetic, thermal, etc.) in a system are governed by the first law of thermodynamics. There has been considerable research activity in *nonelectrically-aided* compaction of powders, for example, see Akisanya et al. [1], Anand and Gu [4], Brown and Abou-Chedid [9], Domas [18], Gu et al. [37], Fleck [28], Tatzel [99] and Zohdi [125]. There has been relatively little analysis of electrically-aided sintering, and it is an area currently under investigation by the author, utilizing the techniques discussed in previous chapters (Zohdi[140]).

9.10 Inverse Problems and Material Design

An important aspect of any model is identification of parameters which force the system behavior to match a (desired) target response. For example, in the ideal case, one would like to determine the type of microstructural parameters that produce certain effective responses, via numerical simulations, in order to guide or minimize time-consuming laboratory tests. A relatively straightforward way of achieving this is to consider inverse problems whereby particulate parameters are sought which deliver a desired overall behavior by minimizing a cost. The identification of microstructural parameters which force the system behavior to match a (desired) target response is essentially a nonconvex inverse problem where the microstructural parameters are the design variables.

For example, a relatively straightforward formulation is to consider inverse problems whereby microstructural parameters are sought which deliver a desired overall behavior by minimizing a cost function, for example,

$$\Pi = w_1 \left(\frac{||\boldsymbol{\epsilon}^* - \boldsymbol{\epsilon}^{*,D}||}{||\boldsymbol{\epsilon}^{*,D}||} \right)^2 + w_2 \left(\frac{||\boldsymbol{\mu}^* - \boldsymbol{\mu}^{*,D}||}{||\boldsymbol{\mu}^{*,D}||} \right)^2 + \text{local constraints}, \tag{9.102}$$

where $\boldsymbol{\epsilon}^{*,D}$ and $\boldsymbol{\mu}^{*,D}$ are desired overall properties, and w_1 and w_2 are design weights that indicate the importance of achieving each component of the objective.

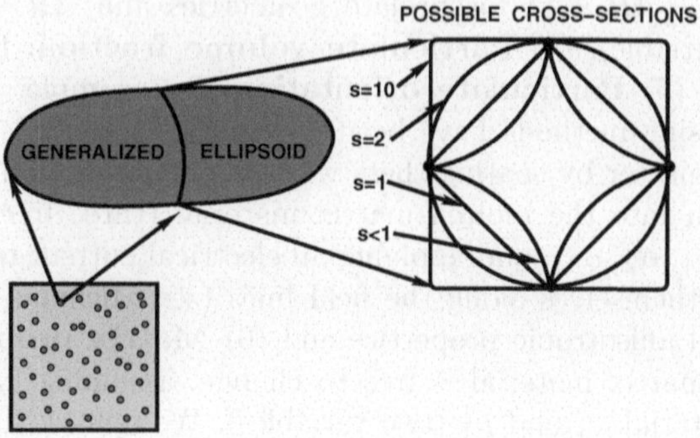

Fig. 9.23. A generalized ellipsoid.

Another example of an objective function that one might minimize is

$$\Pi = w_1 \left(\frac{||\boldsymbol{\epsilon}^* - \boldsymbol{\epsilon}^{*,D}||}{||\boldsymbol{\epsilon}^{*,D}||} \right)^2 + w_2 \left(\frac{||\boldsymbol{\mu}^* - \boldsymbol{\mu}^{*,D}||}{||\boldsymbol{\mu}^{*,D}||} \right)^2 + \text{thermal constraints}. \tag{9.103}$$

Specifically, a microstructural design problem can be set up by defining a N-tuple design vector, denoted $\boldsymbol{\Lambda} \stackrel{\text{def}}{=} (\Lambda_1, \Lambda_2, ..., \Lambda_N)$, for example, consisting of the following components: (1) the properties of the foreign particulate matter (2) the volume fraction of the foreign particulate matter and (3) the topology of the foreign particulate matter. Ellipsoidal shapes are qualitatively useful since the geometry can closely represent a variety of particulate types, for example, platelets when the ellipsoids are oblate or needles (discontinuous fibers) when the ellipsoids are prolate. Such shapes can be generalized by considering the following (Figure 9.23):

$$\left(\frac{|x - x_o|}{r_1}\right)^{s_1} + \left(\frac{|y - y_o|}{r_2}\right)^{s_2} + \left(\frac{|z - z_o|}{r_3}\right)^{s_3} = 1, \tag{9.104}$$

where the s's are exponents. Values of $s < 1$ produce nonconvex shapes, while $s > 2$ values produce "block-like" shapes. The following are free system variables: (1) **Particulate properties:** for example, assuming local isotropy of the particles, ϵ_2 and μ_2 (two variables), (2) **Particulate topology:** for example, the polynomial order of generalized ellipsoids, s (three variables), (3) **Particulate aspect ratio:** for example, defined by $AR \stackrel{\text{def}}{=} \frac{r_1}{r_2} = \frac{r_1}{r_3}$, where $r_2 = r_3$, $AR > 1$ for prolate geometries and $AR < 1$ for oblate shapes (one variable), (4) **Particulate volume fraction:** for example, v_2 (one variable), (5) **Particulate orientation:** for example, within the last decade, processing methods have been developed to control the orientation of particulate matter by coating them with a conducting liquid material, introducing them into the molten matrix material (three free variables, i.e., Euler angles, α_1, α_2, α_3), and applying an electrical current to force the particles to align themselves along the field lines (see Michaud [70]). This will produce overall anisotropic properties and (6) **Matrix properties:** for example, if the matrix material is free to change, assuming local isotropy of the matrix material, ϵ_1 and μ_1 (two variables). We remark that if the particles' orientations are assumed aligned, then three more (angular orientation) parameters can be introduced, $(\alpha_1, \alpha_2, \alpha_3)$. In fact, suspensions can become aligned, for example, along electrical field lines induced by external sources, or due to flow conditions. Thus, the search space grows to more parameters, $\boldsymbol{\Lambda} = (\epsilon_1, \mu_1, \epsilon_2, \mu_2, v_2, s_1, s_2, s_3, AR, \alpha_1, \alpha_2, \alpha_3, etc.)$.

As an example of objective function nonconvexity, consider the approximation of the effective property by using a convex combination of the Hashin–Shtrikman upper and lower effective property bounds

$$\epsilon^{*,\phi} \approx \phi\epsilon^{*,+} + (1 - \phi)\epsilon^{*,-}, \tag{9.105}$$

where the three design variables are ϵ_1, ϵ_2 and v_2. Color-coded plotting this function over the design variable reveals that there are multiple states that achieve the same effective property (Figure 9.24) for the relative permittivity (the same can be generated for the relative permeability and conductivity).

9.10.1 Remarks on Nonconvex Optimization

Generally, the cost function Π will depend on the design variables in a nonconvex and nondifferentiable manner (especially if there are constraints) on

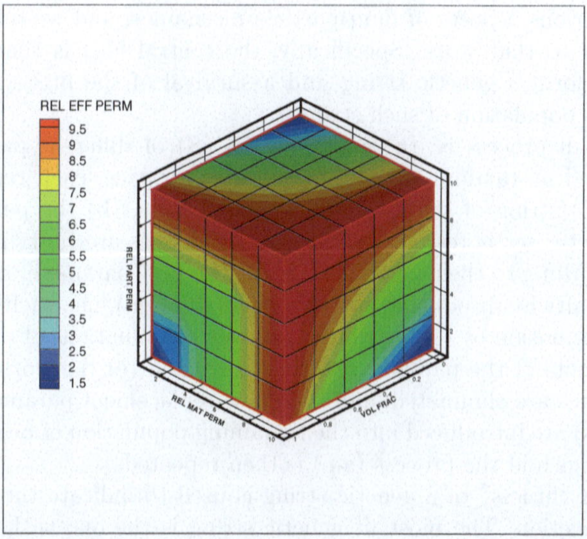

Fig. 9.24. The colors indicate the effective property for the state point:$(v_2, \epsilon_1^r, \epsilon_2^r)$ with $\phi = 0.5$.

the microstructural parameters. One way to minimize such objective functions is by following a two stage approach whereby (1) one determines promising optimal regions in parameter space using (nonderivative) algorithms (such are evolutionary "genetic" algorithms, simulated annealing, etc.) and then (2) applies classical gradient-based schemes in locally convex regions, if the objective functions are smooth enough, since they are generally extremely efficient for the minimization of smooth convex functions. As indicated, the search for convex "pockets" of Π can be achieved by using "genetic" algorithms (GA), before applying classical gradient-based schemes.[9]

Genetic algorithms are search methods based on the principles of natural selection, employing concepts of species evolution, for example, reproduction, mutation and crossover. Implementation typically involves a randomly generated population of fixed-length elemental strings, "genetic" information," each of which represents a specific choice of system parameters. The population of individuals undergoes "mating sequences" and other biologically-inspired events in order to find promising regions of the search space. Such methods can be traced back, at least, to the work of John Holland (Holland [45]). For reviews of such methods, see, for example, Goldberg [34], Davis [13], Onwubiko [76], Kennedy and Eberhart [50] Lagaros et al. [54], Papadrakakis et al. [82–85] and Goldberg and Deb [35]. In Zohdi [124], a genetic algorithm has been developed to treat a wide variety of nonconvex inverse problems

---

[9] An exhaustive review of these methods can be found in the texts of Luenberger [63] and Gill, Murray and Wright [33], while a state of the art can be found in Papadrakakis et al. [86].

involving various aspects of multiparticle mechanics, and we refer the interested reader to that work. Specifically, the central idea is that the system parameters form a genetic string and a survival of the fittest algorithm is applied to a population of such strings.

The overall process is: (a) a population (S) of different parameter sets are generated at random within the parameter space, each represented by a ("genetic") string of the system (N) parameters, (b) the performance of each parameter set is tested, (c) the parameter sets are ranked from top to bottom according to their performance, (d) the best parameter sets (parents) are mated pairwise, producing two offspring (children), i.e., each best pair exchanges information by taking random convex combinations of the parameter set components of the parents' genetic strings and (e) the worst performing genetic strings are eliminated, and then new replacement parameter sets (genetic strings) are introduced into the remaining population of best performing genetic strings and the process (a-e) is then repeated.

The term "fitness" of a genetic string is used to indicate the value of the objective function. The most fit genetic string is the one with the smallest objective function. The retention of the top fit genetic strings from a previous generation (parents) is critical since if the objective functions are highly nonconvex (the present case), there exists a clear possibility that the inferior offspring will replace superior parents. When the top parents are retained, the minimization of the cost function is guaranteed to be monotone (guaranteed improvement) with increasing generations. There is no guarantee of successive improvement if the top parents are not retained, even though nonretention of parents allows more new genetic strings to be evaluated in the next generation. In the scientific literature, numerical studies imply that for sufficiently large populations, the benefits of parent retention outweigh this advantage and any disadvantages of "inbreeding," i.e., a stagnant population. For more details on this so-called "inheritance property," see Davis [13] or Kennedy and Eberhart [50]. In the upcoming algorithm, inbreeding is mitigated since with each new generation, new parameter sets, selected at random within the parameter space, are added to the population. Previous numerical studies of the author (Zohdi [124]) have indicated that not retaining the parents is suboptimal due to the possibility that inferior offspring will replace superior parents. Additionally, parent retention is computationally less expensive since these parameter sets do not have to be reevaluated (or ranked) in the next generation. An implementation of such ideas is as follows (Zohdi [124]):

- **STEP 1:** Randomly generate a population of S starting genetic strings, $\boldsymbol{\Lambda}^i, (i = 1, ..., S)$:
  $\boldsymbol{\Lambda}^i \stackrel{\text{def}}{=} \{\Lambda^i_1, \Lambda^i_2, \Lambda^i_3, \Lambda^i_4 ... \Lambda^i_N\}$
- **STEP 2:** Compute fitness of each string $\Pi(\boldsymbol{\Lambda}^i)$, (i=1, ..., S)
- **STEP 3:** Rank genetic strings: $\boldsymbol{\Lambda}^i$, (i=1, ..., S)
- **STEP 4:** Mate nearest pairs and produce two offspring, (i=1, ..., S)
  $\boldsymbol{\lambda}^i \stackrel{\text{def}}{=} \Phi^{(I)} \boldsymbol{\Lambda}^i + (1 - \Phi^{(I)}) \boldsymbol{\Lambda}^{i+1}, \quad \boldsymbol{\lambda}^{i+1} \stackrel{\text{def}}{=} \Phi^{(II)} \boldsymbol{\Lambda}^i + (1 - \Phi^{(II)}) \boldsymbol{\Lambda}^{i+1}$

- **NOTE:** $\Phi^{(I)}$ and $\Phi^{(II)}$ are random numbers, such that $0 \leq \Phi^{(I)}, \Phi^{(II)} \leq 1$, which are different for each component of each genetic string
- **STEP 5:** Kill off bottom $M < S$ strings and keep top $K < N$ parents and top K offspring (K offspring+K parents+M=S)
- **STEP 6:** Repeat Steps 1-6 with top gene pool (K offspring and K parents), plus M new, randomly generated, strings
- **Option:** Rescale and restart search around best performing parameter set every few generations
- **Remark:** After application of such a global search algorithm, one can apply a gradient-based method, if the objective function is sufficiently smooth in that region of the parameter space. In other words, if one has located a convex portion of the parameter space with a global genetic search, one can employ gradient-based procedures locally to minimize the objective function further since they are generally much more efficient for convex optimization of smooth functions. In such procedures, in order to obtain a new directional step for $\boldsymbol{\Lambda}$, one must solve the following system

$$[\boldsymbol{I\!H}]\{\Delta\boldsymbol{\Lambda}\} = -\{\boldsymbol{g}\}, \tag{9.106}$$

where $[\boldsymbol{I\!H}]$ is the Hessian matrix ($N \times N$), where $\{\Delta\boldsymbol{\Lambda}\}$ is the parameter increment ($N \times 1$), and $\{\boldsymbol{g}\}$ is the gradient ($N \times 1$). We shall not employ this second (postgenetic) stage in this work. An exhaustive review of these methods can be found in the texts of Luenberger [63] and Gill, Murray and Wright [33], while a novel method can be found in Papadrakakis et al. [86].

Similar ideas have been applied to randomly dispersed particulate media with solid binders in Zohdi [124].

Table 9.1. An example of one-dimensional data.

x_i	$f(x_i)$
x_1	$f(x_1)$
x_2	$f(x_2)$
x_3	$f(x_3)$
x_4	$f(x_4)$

9.10.2 Database Approaches

In closing, it is important to mention emerging approaches that are based on interpolation of data comprised of microvariables and measurements of effective responses. For example, consider that a Lagrangian Polynomial of order N is constructed from a set of $N+1$ data points:

$$P_N(x) = \left(\sum_{i=1}^{N} \frac{\Pi_{j=1,j\neq i}^{N}(x - x_j)}{\Pi_{j=1,j\neq i}^{N}(x_i - x_j)}\right) f(x_i) \stackrel{\text{def}}{=} \sum_{i=1}^{N} \Phi_i(x) f(x_i) \tag{9.107}$$

where f_i are the responses at the x_i. This representation does not require even spacing nor that the x values are arranged in any particular order (Table 10.1). The x values must, however, be unique. This is a nodal basis, which has the property that

$$\Phi_i(x_j) = \delta_{ij}, \tag{9.108}$$

and thus

$$P_N(x_i) = f(x_i). \tag{9.109}$$

For multidimensional data, we write

$$P_N(x, y, z, ...) = \left(\sum_{i=1}^{N+1} \frac{\Pi_{j=1,j\neq i}^{N+1}(x - x_j)\ \Pi_{j=1,j\neq i}^{N+1}(y - y_j)\ \Pi_{j=1,j\neq i}^{N+1}(z - z_j)}{\Pi_{j=1,j\neq i}^{N+1}(x_i - x_j)\ \Pi_{j=1,j\neq i}^{N+1}(y_i - y_j)\ \Pi_{j=1,j\neq i}^{N+1}(z_i - z_j)} ... f(x_i, y_i, z_i, ...) \right) \tag{9.110}$$

where, depending on the application, for example, relative permittivity , $(x, y, z) = (v_2, \epsilon_1^r, \epsilon_2^r)$, or for relative permeability, (v_2, μ_1^r, μ_2^r) and for relative conductivity, $(v_2, \sigma_1^r, \sigma_2^r)$. Clearly, the quality of the interpolation will depend on the number of data points. However, with the exponential growth of databases on the Internet, a type of "table lookup," combined with interpolations, such as the one introduced or others, and computational methods to fill in database "holes," may prove in the near future to be invaluable.

10

Concluding Remarks and Emerging Applications in the Biological Sciences

In closing, we discuss emerging applications in bioelectromagnetics to Red Blood Cells (RBCs), which are responsible for the transport of oxygen and carbon dioxide, and are the most prevalent type of cells in human blood. The average cellular volume of each cell (82-96 femtoliters) is occupied by a high concentration of the oxygen carrying protein hemoglobin at a concentration of 30-36 %. The lifespan of the human RBCs is approximately 120 days after they are released from the bone marrow as reticulocytes. Typically 4-6 million RBCs per cubic millimeter occupy 41-52 % of blood volume (hematocrit). The typical biconcave shape of RBCs endows the cell with ideal deformability characteristics. This allows RBCs to efficiently perform their function in small capillaries. Alterations in RBC properties, including shape, volume and membrane characteristics will lead to a decreased lifespan and when not compensated by increased production, a lower volume and anemia. Genetic disorders in cytoskeletal proteins (the cell wall "scaffolding") results in RBC pathologies, such as hereditary spherocytosis and hereditary elliptocytosis (Eber and Lux [22], and Gallagher, [31] and [32]). Deviations in cytosolic and membrane proteins may affect the state of hydration of the cell and thereby its characteristics. Figure 10.1 illustrates some examples of unhealthy cell morphologies. In normal blood, echinocytes and stomatocytes can be observed, in addition to discocytes. Acanthocytes are observed in acquired hepatic syndromes, codocytes are found in thalassemia. In sickle cell disease, hemoglobin polymers will distort the shape of the cell and drepanocytes are observed. Elliptocytes are the result of membrane disorders where interactions in the horizontal direction (e.g., spectrin-spectin interactions) are disrupted, and spherocytes are observed in membrane disorders where the interaction between the lipid bilayer and the underlying membrane skeleton is dysfunctional. The number of humans that are affected by Sickle cell disease, Thalassemia and other hemoglobinopathies, runs in the millions (Forget and Cohen [29] and Steinberg et al. [97]). Such disorders lead to altered hemoglobin and result in changes in RBC properties, which is related to blood pathology, including anemia.

T.I. Zohdi: Electromagnetic Properties of Multiphase Dielectrics, LNACM 64, pp. 151–166.
springerlink.com

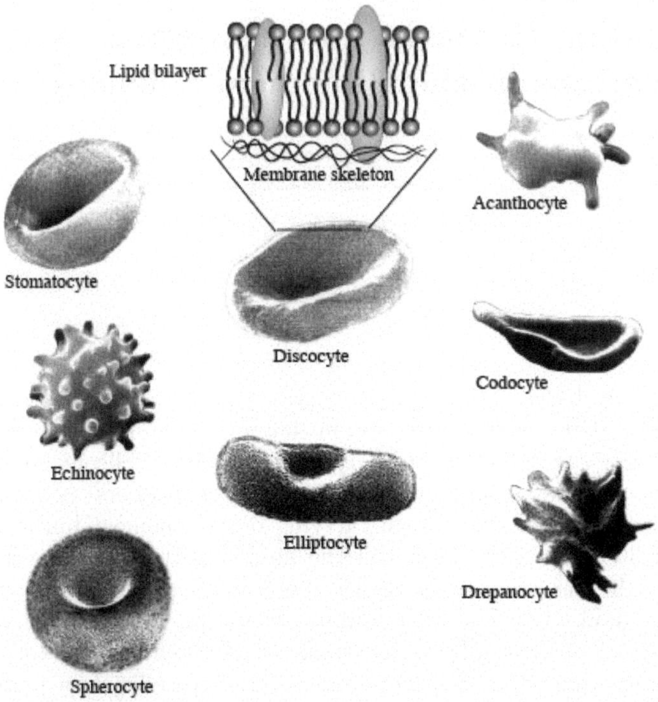

Fig. 10.1. The red blood cell membrane encloses the cytosol with hemoglobin. The membrane consists of a lipid bilayer which interacts with a spectrin skeleton giving the normal red blood cell its typical discocyte shape. The diameter of a healthy red blood cell (7.4-8.2 microns) can be markedly different in disease (2-11 microns) and can exhibit a variety of shapes. The drawings were created by F. Kuypers (see Zohdi, Kuypers and Lee [136]), and based on the photographs of Bessis [6].

10.1 Objectives

Here we follow an approach found in Zohdi, Kuypers and Lee [136]. The rapid testing of RBC data of individuals to define hematologic characteristics and possibly pathology is important. Quick and inexpensive electromagnetic measurements are considered to be an important technique to measure blood properties. The goal of this work is to develop estimates of the volume fraction levels of cells, denoted by v_i, in whole blood from a simple macroscopic electromagnetic measurement of the overall blood properties, at relatively low frequencies (below one GHz). The approach taken here is to develop bounds on the possible volume fraction by inverting the classical bounds on the overall permittivity of electromagnetic mixtures. The usefulness of the approach is that, given the permittivities of the buffer (representing the plasma, known), cell (known) and whole mixture (measured), one can determine the cell volume fraction. This information is useful to compare against the nominal volume fraction (hematocrit) values for healthy individuals.

Remark: For rapid simulation of (high-frequency) light scattering by RBCs, see Zohdi and Kuypers [135].

10.2 Laboratory Experiments

Following Zohdi, Kuypers and Lee [136], The relative permittivity of RBC suspensions was measured at different volume fractions in the radio frequency (RF) range (0.3MHz-1GHz) using an open-ended coaxial line sensor technique (Grant [36]). The relative permittivity of RBC suspensions was corrected to compensate for electrode polarization effects, followed by curve-fitting which enabled extrapolations to calculate the static relative permittivity.

Preparation of human red blood cells (RBC): Blood samples from healthy donors were collected in EDTA as anticoagulant, after informed consent, at the Childrens Hospital Oakland Research Institute (CHORI). Whole blood was kept at four degrees C and used within 24 hours after collection. RBCs were isolated by centrifugation, plasma and the buffy coat were removed and the RBCs were washed three times in ten volumes of phosphate-buffered saline. The RBCs were resuspended at approximately 30 % volume fraction in phosphate buffered saline and stored at four degrees C until used within 48 hours. The exact cell count in the suspension was determined using the ADIVA 120 hematology system, and the suspensions were diluted to the indicated volume fraction levels.

Experimental apparatus for permittivity measurements: We used an open-ended coaxial line sensor (Grant [36]) made by an SMB jack (male) connector and an E5071B network analyzer (Agilent Technology). The network analyzer measured the sensor admittance, whose imaginary part represented the relative permittivity of a surrounding material at the sensor end. In order to standardize the sensor, we used two kinds of dielectric reference materials; air and deionized water. The air measurement corrected the RF phase offset between calibration and sensor planes. The deionized water measurement (relative permittivity of 78.3 (Grant [36])) was used as a reference value of the relative permittivity. After the sensor standardization, the relative permittivity of a buffer solution and RBC suspensions at indicated volume fractions were measured in the frequency range from 0.3MHz to 1GHz at 201 distinct values.

Correction for the electrode polarization effect: The relative permittivity values contained errors from the electrode polarization effect (Schwan [93] and [94]). At RBC concentrations of up to 10 % volume fraction, the electrode polarization effect of RBC suspensions is the same as that of the buffer solution. In our experiments we diluted the RBC suspensions within a $2-10\%$ range, as at higher concentrations, the electrode polarization effect shows complex and nonlinear behavior that can only be characterized using special experimental methods (reviewed by Schwan [93]), and is beyond the present scope of this work. Figure 10.2 shows the corrected relative permittivity of RBC suspensions with different concentration levels. Our measured permittivity data correlates well with previous reports (Schwan [94]).

Relative permittivity: In order to obtain the static relative permittivity, we used curve fittings and extrapolations from the corrected relative permittivity with sigmoidal curve functions used by Schwan [94]. The solid lines in Figure 10.2 are fitted curves from the measured values, and allowed *extrapolation to the static relative permittivity* from the curves. Figure 10.2 shows the averages and standard deviations of the static relative permittivity from five repeated measurements.

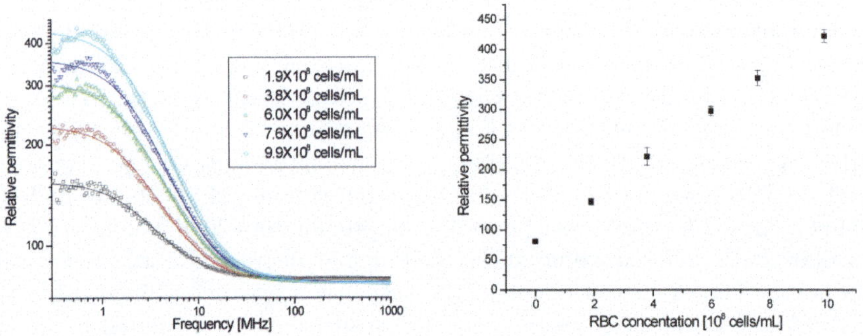

Fig. 10.2. Measured relative permittivity of RBC suspensions for different cell concentrations ($0.019 \leq v_2 \leq 0.099$), which directly correspond to volume fractions occupied by the cells: (a) Measured relative permittivity in the frequency range from 0.3MHz to 1GHz. Dots and solid lines represent measured data points and their fitted curves (solid lines on the left), respectively; (b) Static relative permittivity for different cell concentrations (Zohdi, Kuypers and Lee [136]).

10.3 Theoretical Estimates: Extraction of Cell Data from Cell-in-Solution Measurements

Within a sample of material, a mixture of cells and buffer in the case of interest, the properties are characterized by a spatially variable permittivity $\epsilon(\boldsymbol{x})$.

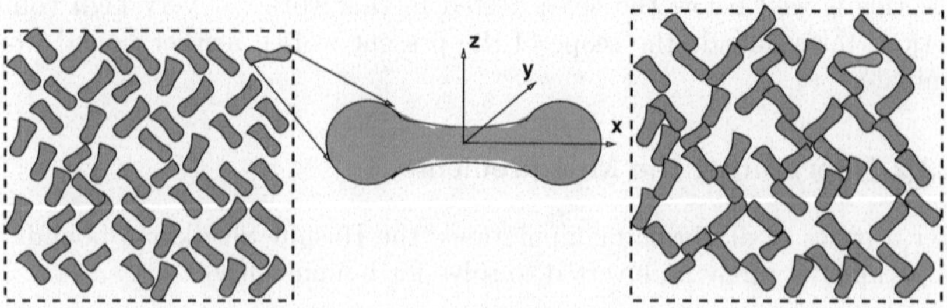

Fig. 10.3. Left: A representative volume element (RVE), with well-separated cells. Right: An RVE with a chain-like network of cells for the same volume fraction as on the left (Zohdi, Kuypers and Lee [136]).

Typically, as for any other heterogeneous material, in order to characterize the overall effective permittivity, a relation between averages is determined, that is,

$$\langle \boldsymbol{D} \rangle_{\Omega} = \boldsymbol{\epsilon}^{*} \cdot \langle \boldsymbol{E} \rangle_{\Omega}, \tag{10.1}$$

where $\langle \cdot \rangle_{\Omega} \stackrel{\text{def}}{=} \frac{1}{|\Omega|} \int_{\Omega} \cdot \, d\Omega$ is the averaging operator and $\boldsymbol{D}$ and $\boldsymbol{E}$ are the electrical flux and electric field within a statistically representative volume element (RVE) of volume $|\Omega|$. The permittivity tensor $\boldsymbol{\epsilon}^{*}$ represents the effective overall permittivity of the cell and buffer mixture.

10.3.1 Effective Permittivity Estimates

We shall use the well-known Hashin–Shtrikman bounds [42], which are the tightest possible on isotropic effective responses, with isotropic two phase microstructures, where only the volume fractions and phase contrasts of the constituents are known. Note that no further geometric information, such as, the number and nature of cells, etc., contributes to these bounds. Generally, the upper bound is more accurate (closer to the actual effective property) for a microstructure comprised of a high permittivity interstitial network/matrix (buffer) surrounding well-separated low permittivity regions. If the cells touch one another, then the upper bound is quite accurate since the cells form a high permittivity network that separates low-permittivity buffer regions from one another (Figure 10.3). We shall employ these bounds for the rest of the analysis, however, other methods could also be used. The lower bound is typically more accurate for microstructures where well-separated, high-permittivity, particles (cells) are surrounded by a lower permittivity matrix (buffer). Thus, the lower bound would be more accurate at extremely low volume fraction of cells; this is well below the levels tested in this work. At very high volume fractions (also outside the scope of the present work), neither bound would dominate.

10.3.2 Interpreting the Measurements

After a series of algebraic manipulations, the Hashin–Shtrikman bounds for the permittivity may be inverted to solve for bounds on v_2

$$\underbrace{\frac{(\epsilon^* - \epsilon_1)3\epsilon_2}{(\epsilon^* + 2\epsilon_2)(\epsilon_1 - \epsilon_2)}}_{\stackrel{\text{def}}{=} v_2^-} \leq v_2 \leq \underbrace{\frac{(\epsilon^* - \epsilon_1)(2\epsilon_1 + \epsilon_2)}{(2\epsilon_1 + \epsilon^*)(\epsilon_2 - \epsilon_1)}}_{\stackrel{\text{def}}{=} v_2^+}, \tag{10.2}$$

where inverting the upper bound Hashin–Shtrikman bound generates v_2^- and inverting the lower bound Hashin–Shtrikman bound generates v_2^+. The utility of this inversion is that, given the permittivities of the buffer (known), cell (known) and whole mixture (measured), one can determine the cell volume fraction and compare it to the nominal response of a healthy group of cells. The relation in Equation 10.2 is rather general, and can be used for any cell-in-solution system.

Table 10.1. The measured (relative permittivity, $\epsilon_i = \epsilon_{ri}\epsilon_o$, ϵ_o being the vacuum permittivity) response for radiation at 1 MHz. Volume fractions above 0.1 were difficult to measure because the polarization effect cannot be canceled at the higher volume fraction values. The relative permittivity of the buffer alone was measured to be $\epsilon_{r1} = 81.12$.

v_2	ϵ_r^*(measured)	v_2^-(from ϵ^{*+})	v_2^+(from ϵ^{*-})	θ_v(calculated)	$\epsilon_r^{*-}(v_2)$	$\epsilon_r^{*+}(v_2)$	θ_ϵ(calculated)
0.019	147.14	0.0177382	0.2348342	0.0058123	85.6437922	157.0738288	0.9289679
0.038	221.97	0.0350468	0.3816693	0.0085199	90.3389545	233.9887411	0.9163330
0.060	297.05	0.0534005	0.4894688	0.0151342	96.0026452	324.2729706	0.8807423
0.076	352.67	0.0668532	0.5490572	0.0189688	100.2838741	390.7757183	0.8688235
0.099	422.40	0.0835484	0.6077450	0.0294767	106.6929048	487.6406536	0.8287412

In order to illustrate how the inverted bounds may be used to aid in interpreting experiments, consider that the relative (to vacuum) permittivity of hemoglobin (or "cell-only") responses is approximately $\epsilon_2/\epsilon_0 \approx 6000$ (an estimate, since it is difficult to measure a "pure" interior cell material), while the buffer was $\epsilon_1/\epsilon_0 = 81.12$ (measured). With this information, and the overall response measurements made before, one can generate the volume fraction bounds (Table 10.1). Recall, however, that the upper bound is quite accurate for touching cells, forming networks, which appears to be the present case. In order to provide a clear, continuous, calibration, we form a convex combination of the upper and lower bounds ($0 \leq \theta_v \leq 1$)

$$v_2 = \theta_v v_2^+ + (1 - \theta_v) v_2^- , \tag{10.3}$$

where, provided measurements are made for v_2 (as they have been made in this study), one can solve for the θ_v value, representing which bound (and microstructural morphology) is dominant, namely,

$$\theta_v = \frac{v_2 - v_2^-}{v_2^+ - v_2^-} . \tag{10.4}$$

Specifically,

- $\theta_v < 1/2$ indicates that v_2^- is dominant, which is generated by the Hashin–Shtrikman *upper* bound, where the effective response is indicative of cells touching, while
- $\theta_v > 1/2$ indicates that v_2^+ is dominant, which is generated by the Hashin–Shtrikman *lower* bound, where the effective response is indicative of well-separated cells.

The derived bounds on v_2 ($v_2^- \leq v_2 \leq v_2^+$) are general and hold for any combination of known values of ϵ_1, ϵ_2 and ϵ^*. For the case at hand, θ_v (Table 10.1) is comfortably less than 1/2 (the lower volume fraction bound being more accurate, generated by the Hashin–Shtrikman permittivity upper bound) leading to the hypothesis that the cells are touching and forming high-permittivity networks. Although the high-permittivity network effect is dominant at very low volume fractions, the effect is steadily less so with increasing volume fraction, characterized by an increasingly larger θ_v. Alternatively, one could directly relate the quality (narrowness) of the bounds on the permittivity to the measured value by writing ($0 \leq \theta_\epsilon \leq 1$)

$$\epsilon^* = \theta_\epsilon \epsilon^{*+} + (1 - \theta_\epsilon) \epsilon^{*-} , \tag{10.5}$$

where, provided measurements are made for ϵ^* (as they have been made in this study), one can solve for the θ_ϵ value, representing which bound (and microstructural morphology) is dominant, that is,

$$\theta_\epsilon = \frac{\epsilon^* - \epsilon^{*-}}{\epsilon^{*+} - \epsilon^{*-}} . \tag{10.6}$$

As with the θ_v metric (however, essentially reversed): (1) if $\theta_\epsilon > 1/2$, this indicates that ϵ^{*+} is dominant, which is the Hashin–Shtrikman *upper* bound, where the effective response is indicative of cells touching, while (2) if $\theta_\epsilon < 1/2$ this indicates that ϵ^{*-} is dominant, which is the Hashin–Shtrikman *lower* bound, where the effective response is indicative of well-separated cells. This is an alternative, yet equivalent representation of the cell-system behavior. As Table 10.1 indicates, θ_ϵ is comfortably larger than 1/2, again indicating that the Hashin–Shtrikman upper bound is dominant.

Remark: It is important to note that the hypothesized network connectivity of cells occurs at low volume fractions, as tested in our experiments, where the cell suspensions were continuously mixed to avoid cells from settling at the bottom of the test apparatus. We emphasize that, in the experiments, the mean values of different random realizations of cells were measured and recorded in the previous Figure 10.2 and Table 10.1. It is important to emphasize that this connectivity is not a Rouleux formation ("coin stacking") arrangement which is sometimes observed at high volume fractions.

10.3.3 Observations

The correlation of the measurements and the analytical expressions (bounds) imply that RBCs form high-permittivity cell-networks by making cell-to-cell contact, even at low volume fractions. This observation is extremely important in order to make reliable estimates of volume fraction levels from overall cell-in-solution measurements. However, in order to strengthen this cell-network hypothesis, we continue with large-scale numerical simulations.

10.4 Computational Simulation of Multiple Cell Samples

10.4.1 Outline of the Approach

Direct computational experiments are achieved by solving Maxwell's equations numerically over samples of cells. This type of calculation has been performed for other general thermally-coupled electromagnetic heterogeneous media in Zohdi [137] and [139], and we follow a simplified version of that approach presently. In order to achieve this, we utilize a mathematical representation of the nominal biconcave RBC shape (Figures 10.3 and 10.4)

$$F \stackrel{\text{def}}{=} \left(\frac{2(z-z_o)}{b}\right)^2 - \left(1 - \frac{(x-x_o)^2+(y-y_o)^2}{b^2}\right) \times \left(c_o + c_1\left(\frac{(x-x_o)^2+(y-y_o)^2}{b^2}\right) + c_2\left(\frac{(x-x_o)^2+(y-y_o)^2}{b^2}\right)^2\right)^2 = 0, \tag{10.7}$$

where the geometrical parameters, c_o, c_1 and c_2 are found in the work of Evans and Fung [23] (for example, $c_o = 0.207161$, $c_1 = 2.002558$ and $c_2 = -1.122762$), and where b is the radius of the cells. One can use this parametrization, in conjunction with a finite-difference mesh or finite-element mesh, to develop a representation of a group of cells. For the numerical experiments, the position and orientation of each cell in the sample was random and not touching any neighboring cells (Figures 10.4 and 10.5). The classical random sequential addition (RSA) algorithm (for placing nonintersecting objects in a domain) was used to place nonoverlapping cells randomly into the domain of interest (Widom [108]). This algorithm was adequate for the volume fraction range of interest. However, if higher volume fractions are desired, more sophisticated algorithms, for example, the well-known, equilibrium-based, Metropolis algorithm can be used, while for extremely high volume fractions, approaches based on particle flow and growth are preferable (Torquato [104], Kansaal et al. [49] and Donev et al. [19–21]). The random orientation of the cells was controlled by using a random angle in a standard rotational coordinate transformation of the axes in Equation 10.7 after placement within the domain.

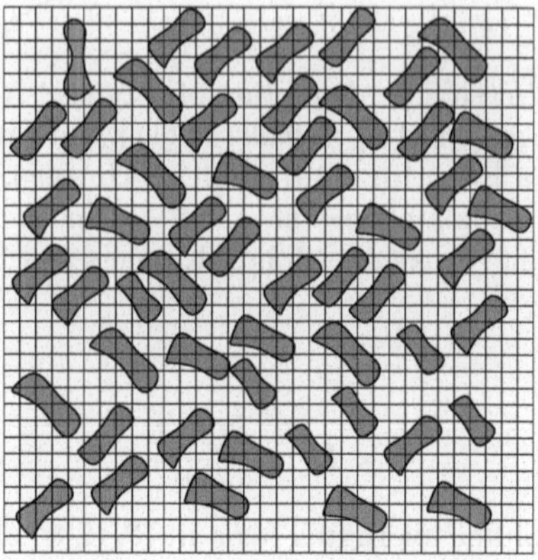

Fig. 10.4. A schematic of a mesh (actual meshes are much finer, see Figure 10.5). (Zohdi, Kuypers and Lee [136]).

10.4.2 Computational Effective Property Calculation

We now develop a direct numerical scheme based on the Finite Difference Time Domain (FDTD) Method to determine the electromagnetic response of a sample of blood by solving the coupled Maxwellian system (Faraday's law)

$$\nabla \times \boldsymbol{E} = -\frac{\partial(\boldsymbol{\mu} \cdot \boldsymbol{H})}{\partial t} \tag{10.8}$$

and (Ampere's law)

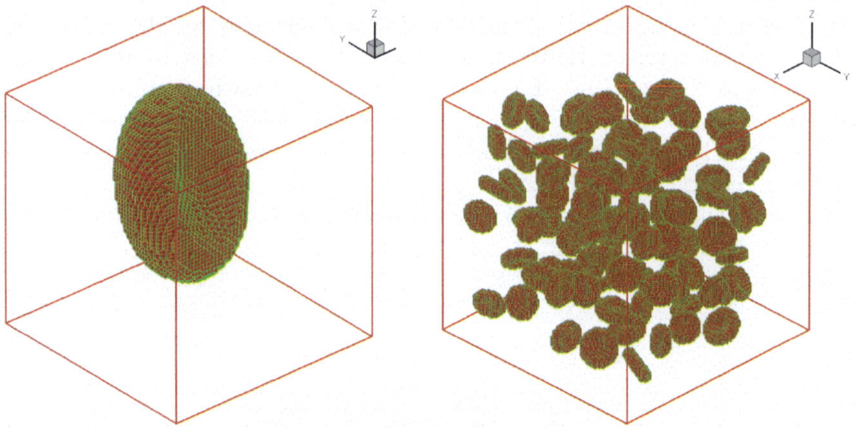

Fig. 10.5. Numerical (mesh) representation of a typical RBC (left) and a sample of blood at a volume fraction of $v_2 = 0.094$ (right). On the right is the resolution of the cell system with a $101 \times 101 \times 101$ mesh which has 6,181,806 *electromagnetic degrees of freedom*. Approximately beyond the 61/81 mesh-density level, there were no perceivable changes in the results (Zohdi, Kuypers and Lee [136]).

$$\nabla \times \boldsymbol{H} = \frac{\partial(\boldsymbol{\epsilon} \cdot \boldsymbol{E})}{\partial t}. \tag{10.9}$$

In order to perform volume averaging over an RVE, to obtain $\boldsymbol{\epsilon}^*$ computationally, we employ a "framing" technique, whereby uniform far-fields are applied on the boundary of a large sample, and the fields are averaged over an interior subsample, in order to avoid boundary layer effects which occur from imposing the uniform fields on the larger sample exterior (Figure 10.6). This is akin to exploiting a "St. Venant-type" decay effect, commonly exploited in solid mechanics, to avoid boundary layers. The approach provides a way of determining what the microstructure really experiences, without "bias" from the boundary loading. In order to determine $\boldsymbol{\epsilon}^*$ (or any effective property), one specifies three uniform (spatially constant) linearly independent loadings of the form, for example, either $\boldsymbol{E}$ or $\boldsymbol{D}$. Each of the loadings provides three linearly independent equations which can be used to solve for the nine constants in a general anisotropic effective permittivity tensor $\boldsymbol{\epsilon}^*$ (in reality, there are only six constants because $\boldsymbol{\epsilon}^*$ is symmetric). If the overall response is isotropic, which is the present case since we have a mixture of randomly oriented cells, then only one test loading is needed. As before, the practical implementation of "framing" is:

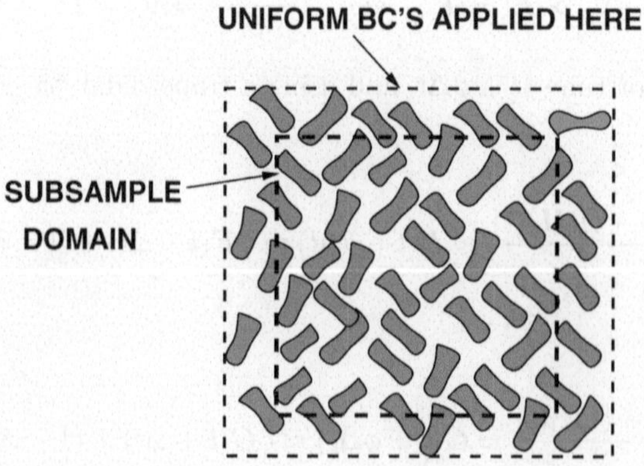

Fig. 10.6. A cell sample with uniform boundary fields applied to its exterior and an interior subsample for averaging purposes (to avoid boundary layer effects) Zohdi, Kuypers and Lee [136].

- *Step (1):* Generate a sample with a certain number of cells in its interior to meet the volume fraction under investigation,
- *Step (2):* For the effective property calculation (averaging), select a subsample ("a sub-box," Figure 10.6) in the interior (to avoid boundary layer effects that arise from the imposition of uniform boundary conditions),
- *Step (3):* Repeat Steps (1) and (2) for different random realizations for at a given sample sample size, and ensemble average the effective properties to determine a mean (ϵ^*) value,
- *Step (4):* Repeat Steps (1)-(4) for a larger sample,
- *Step (5):* Continue the process (Steps (1)-(4)) until the effective property ceases to change to within an acceptable tolerance.

The general discretization process follows an approach found in Zohdi [137] and [139], and is briefly outlined in the next section.

Remark: There are similar relations holding for the magnetic permeability, relating the magnetic field flux ($\boldsymbol{B}$) and the magnetic field ($\boldsymbol{H}$), $\langle \boldsymbol{B} \rangle_\Omega = \boldsymbol{\mu}^* \cdot \langle \boldsymbol{H} \rangle_\Omega$, and in the case of overall isotropy

$$\mu^* \stackrel{\text{def}}{=} \sqrt{\frac{\langle \boldsymbol{B} \rangle_\Omega \cdot \langle \boldsymbol{B} \rangle_\Omega}{\langle \boldsymbol{H} \rangle_\Omega \cdot \langle \boldsymbol{H} \rangle_\Omega}}. \tag{10.10}$$

10.4.3 Numerical Discretization of Maxwell's Equations

Following a general procedure introduced earlier, let us now consider the direct numerical simulation of a blood sample by defining

$$\frac{\partial(\mu \boldsymbol{H})}{\partial t} = -\nabla_x \times \boldsymbol{E} \stackrel{\text{def}}{=} \boldsymbol{F} \quad \text{and} \quad \frac{\partial(\epsilon \boldsymbol{E})}{\partial t} = \nabla_x \times \boldsymbol{H} \stackrel{\text{def}}{=} \boldsymbol{G}. \tag{10.11}$$

We discretize for time=$t + \phi \Delta t$, and using a trapezoidal "$\phi - scheme$" ($0 \leq \phi \leq 1$),

$$\frac{(\mu \boldsymbol{H})(t+\Delta t) - (\mu \boldsymbol{H})(t)}{\Delta t} \approx \boldsymbol{F}(t+\phi\Delta t) \approx \phi \boldsymbol{F}(t+\Delta t) + (1-\phi)\boldsymbol{F}(t) \tag{10.12}$$

and

$$\frac{(\epsilon \boldsymbol{E})(t+\Delta t) - (\epsilon \boldsymbol{E})(t)}{\Delta t} \approx \boldsymbol{G}(t+\phi\Delta t) \approx \phi \boldsymbol{G}(t+\Delta t) + (1-\phi)\boldsymbol{G}(t). \tag{10.13}$$

Rearranging yields

$$\boldsymbol{H}(t+\Delta t) \approx \frac{(\mu \boldsymbol{H})(t)}{\mu(t+\Delta t)} + \frac{\Delta t}{\mu(t+\Delta t)} (\phi \boldsymbol{F}(t+\Delta t) + (1-\phi)\boldsymbol{F}(t)) \tag{10.14}$$

and

$$\boldsymbol{E}(t+\Delta t) \approx \frac{(\epsilon \boldsymbol{E})(t)}{\epsilon(t+\Delta t)} + \frac{\Delta t}{\epsilon(t+\Delta t)} (\phi \boldsymbol{G}(t+\Delta t) + (1-\phi)\boldsymbol{G}(t)). \tag{10.15}$$

Numerically, the components of the curl of functions such as $\boldsymbol{E}$ are approximated by central finite difference stencils of the form (Figure 10.4):

$$\frac{\partial \boldsymbol{E}(x)}{\partial x} \approx \frac{\boldsymbol{E}(x+\Delta x) - \boldsymbol{E}(x-\Delta x)}{2\Delta x}, \ etc. \tag{10.16}$$

In order to construct a solution, the algorithm is as follows:

- (1) *Spatiotemporal discretization:* Construct spatial derivative terms such as

$$\frac{\partial \boldsymbol{E}(x)}{\partial x} \approx \frac{\boldsymbol{E}(x+\Delta x) - \boldsymbol{E}(x-\Delta x)}{2\Delta x}, \tag{10.17}$$

  for each node (i, j, k), in conjunction with time-discretizations in Equations 10.14 and 10.15, leading to coupled systems abstractly written as

$$\boldsymbol{H}^{t+\Delta t} = \mathcal{F}(\boldsymbol{E}^{t+\Delta t}, \boldsymbol{H}^{t+\Delta t}) \quad \text{and} \quad \boldsymbol{E}^{t+\Delta t} = \mathcal{G}(\boldsymbol{E}^{t+\Delta t}, \boldsymbol{H}^{t+\Delta t}). \tag{10.18}$$

- (2) *Staggered solution:* Compute $\boldsymbol{E}$-field with $\boldsymbol{H}$ fixed, then compute $\boldsymbol{H}$-field with $\boldsymbol{E}$ fixed, and iterate, $K = 1, 2, ...$ for

$$\boldsymbol{H}^{t+\Delta t,K} = \mathcal{F}(\boldsymbol{E}^{t+\Delta t,K}, \boldsymbol{H}^{t+\Delta t,K-1}) \quad \text{and} \quad \boldsymbol{E}^{t+\Delta t,K} = \mathcal{G}(\boldsymbol{E}^{t+\Delta t,K-1}, \boldsymbol{H}^{t+\Delta t,K-1}). \tag{10.19}$$

- (3) *Staggering error check:* Compute error measures: $\varpi^*_K \stackrel{\text{def}}{=} max(\varpi_{E-K}, \varpi_{H-K}),\ i = 1, ..., nodes$

$$\varpi_{K-E} \stackrel{\text{def}}{=} \frac{\sum_{i=1}^{N} ||\boldsymbol{E}_i^{L+1,K} - \boldsymbol{E}_i^{L+1,K-1}||}{\sum_{i=1}^{N} ||\boldsymbol{E}_i^{L+1,K} - \boldsymbol{E}_i^{L}||}, \quad \varpi_{K-H} \stackrel{\text{def}}{=} \frac{\sum_{i=1}^{N} ||\boldsymbol{H}_i^{L+1,K} - \boldsymbol{H}_i^{L+1,K-1}||}{\sum_{i=1}^{N} ||\boldsymbol{H}_i^{L+1,K} - \boldsymbol{H}_i^{L}||}. \tag{10.20}$$

- (4) *Updates:* When the tolerance is met, $\varpi^* \leq C_{tol}$, then increment time forward: $t = t + \Delta t$.

Remarks: The time step size used in the upcoming simulations allowed the algorithm to easily converge within a few iterations at each time step. However, for more strongly coupled systems, temporal (time step) adaptivity is usually needed (as previously noted earlier in the monograph). Implicit, iterative procedures, with time step adaptivity, are ideal for strongly-coupled thermoelectromagnetic and nonlinear systems.

10.4.4 A Model Problem

As a model problem, we considered a group of N_p randomly dispersed, well-separated cells, of equal size, in a cubical domain of dimensions, $D \times D \times D$ ($0.80 \times D$, was selected as the length-scale of the subsample). This sample size (and hence the number of cells, with the volume fraction held constant) was successively enlarged until there were no significant changes in the overall system response (ϵ^*) for further enlargements. For a more in-depth discussion on size-effect issues for this class of systems, see the works of Zohdi [120–128]. We found that a sample that contained approximately $N_p = 100$ cells (Figure 10.5) was adequate, i.e., further sample enlargements produced negligible changes in the computed effective properties. The following system parameters were used: $\epsilon_o = 8.854 \times 10^{-12}$ farads/meter, $\mu_o = 4\pi \times 10^{-7} Ns^2/C^2$, the electric field on the boundary (linearly-growing), $\boldsymbol{E}|_\Omega = (10^9, 10^9, 10^9)\frac{t}{T}$, initial conditions, $\boldsymbol{E}(t = 0) = (0, 0, 0)$, magnetic field on the boundary (linearly-growing) and $\boldsymbol{H}|_\Omega = (10^9, 10^9, 10^9)\sqrt{\frac{\mu_o}{\epsilon_o}}\frac{t}{T}$, initial conditions, $\boldsymbol{H}(t = 0) = (0, 0, 0)$. The relative magnetic permeability was set to one for the cells, $\mu_{2r} = 1$ and for the matrix, $\mu_{1r} = 1$ (the magnetic heterogeneity (mismatch) is essentially negligible). The relative permittivity of the cells was approximately $\epsilon_2/\epsilon_0 = 6000$, while the buffer was $\epsilon_1/\epsilon_0 = 81.12$.

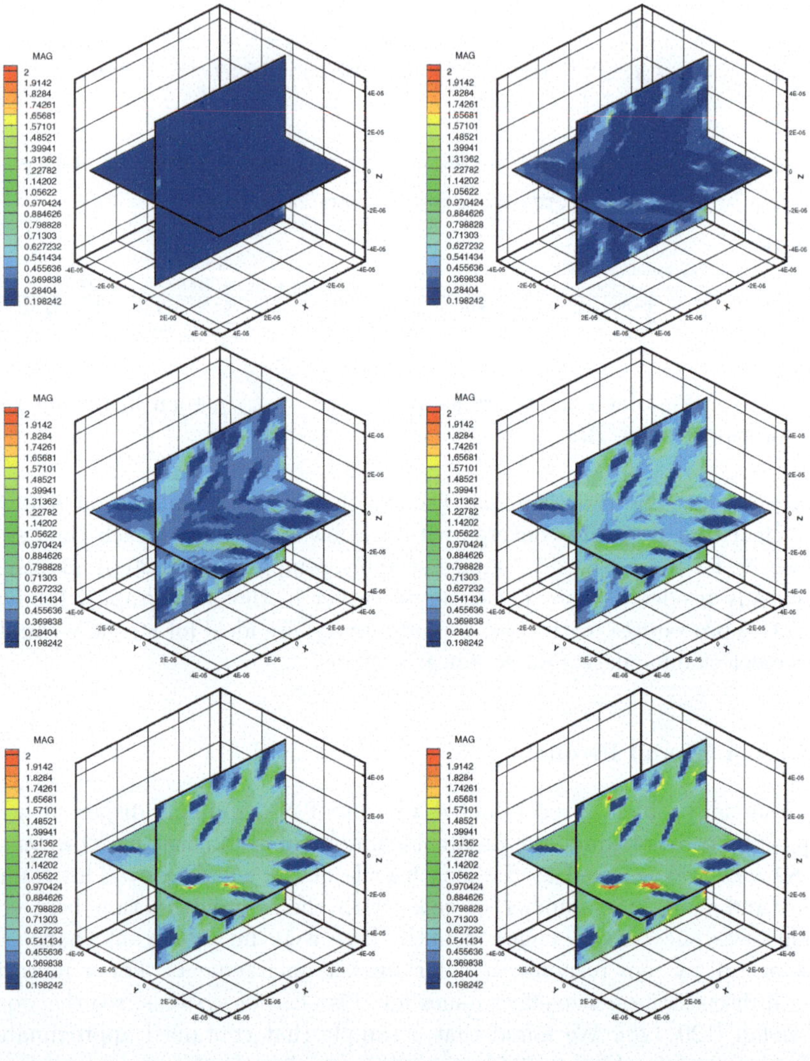

Fig. 10.7. Starting from left to right and top to bottom, the progressive evolution of the (normalized) electrical field $\frac{||\boldsymbol{E}||}{||\boldsymbol{E}|_{\partial\Omega(t=T)}||}$ magnitude within an 0.8-probe sub-sample (20 % into the interior). $\boldsymbol{E}|_{\partial\Omega} = (10^9, 10^9, 10^9)\frac{t}{T}$ is the field applied on the exterior surface of the sample of blood at a volume fraction of $v_2 = 0.094$ (Zohdi, Kuypers and Lee [136]).

The cell size and volume fraction were determined by a cell/sample size ratio, which was defined via a subvolume size $V \stackrel{\text{def}}{=} \frac{D\times D\times D}{N_p}$. As an example, we used

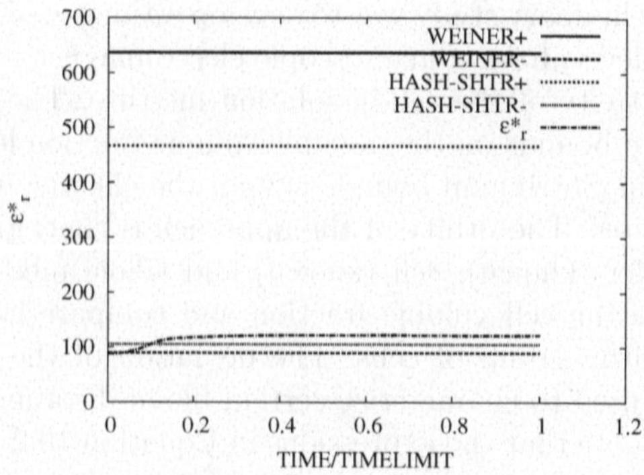

Fig. 10.8. The progressive evolution of the effective permittivity, ϵ^*, from direct numerical simulation, within an 0.8-probe subsample (20 % into the interior). $\boldsymbol{E}|_{\Omega} = (10^9, 10^9, 10^9)\frac{t}{T}$ is the field applied on the exterior surface of the sample of blood at a volume fraction of $v_2 = 0.094$. The steady state value for this well-separated system was approximately $\epsilon_r^* \approx 125$, as opposed to the measured value (for the same volume fraction) of approximately $\epsilon_r^* \approx 400$, leading to the hypothesis that the real system must possess cell-network chains.

a volume fraction of approximately $v_2 = 0.094$, which corresponded to cells with a length scale of $\mathcal{L} = 0.375$, which is a nondimensional ratio between the major cell radius (b) and the subvolume given by $\mathcal{L} \stackrel{\text{def}}{=} \frac{b}{V^{\frac{1}{3}}}$. The length-scale dimension of the sample was D=50 microns.

The meshes were sequentially refined in the following manner: (I) a 41 × 41 × 41 mesh which has 413,526 *nodal electromagnetic unknowns*, (II) a 61 × 61 × 61 mesh which has 1,361,886 *nodal electromagnetic unknowns*, (III) an 81 × 81 × 81 mesh which has 3,188,646 *nodal electromagnetic unknowns* and (IV) a 101 × 101 × 101 mesh which has 6,181,806 *nodal electromagnetic unknowns*. Approximately beyond the 61/81-level, there were no perceivable changes in the results. In Figure 10.8, the progressive evolution of the effective permittivity, ϵ^*, from direct numerical simulation, within an 0.8-probe subsample (20 % into the interior). The steady state relative effective permittivity value for this well-separated system was approximately $\epsilon_r^* = \epsilon^*/\epsilon_o \approx 125$, as opposed to the measured value (for the same volume fraction) of approximately $\epsilon_r^* = \epsilon^*/\epsilon_o \approx 400$. Thus, a system of well-separated cells is incapable of producing the observed experimental results, again suggesting the hypothesis that the the real system possesses a network of touching cells.

10.5 Discussion and Concluding Remarks

The goal of this preliminary study was to develop estimates of the volume fraction (hematocrit) levels from a macroscopic electromagnetic measurement of the overall permittivity of the cell-in-solution mixture. The approach taken here was to develop bounds on the possible volume fraction level by inverting the classical Hashin–Shtrikman bounds [42] on the effective response of electromagnetic mixtures. The utility of the approach is that, given the permittivities of the buffer (known), cell (known) and whole mixture (measured), one can determine the cell volume fraction and compare it to the nominal response of a healthy group of cells. The deviation of the electromagnetic properties can be used to characterize certain blood disorders (thalassemia). It is important to note that the expressions in Equation 10.2 are valid for any combination of known values, ϵ_1, ϵ_2 and ϵ^* to obtain bounds on cell volume fraction (v_2) not necessarily just those of RBC mixtures. Through correlation of our laboratory measurements and the analytical expressions (bounds), it is hypothesized that RBCs may form high-permittivity cell-networks by making cell-to-cell contact, even at low volume fractions, which is a subject of ongoing investigation by the author.

11

References

1. Akisanya, A.R., Cocks, A.C.F., Fleck, N.A.: The yield behavior of metal powders. Int. J. Mech. Sci. 39, 1315–1324 (1997)
2. Amdahl, G.: The validity of a single processor approach to achieving large-scale computing capabilities. In: Proceedings of AFIPS Spring Joint Computer Conference, pp. 483–485. AFIPS Press, Atlantic City (1967)
3. Ames, W.F.: Numerical methods for partial differential equations, 2nd edn. Academic Press (1977)
4. Anand, L., Gu, C.: Granular materials: constitutive equations and shear localization. J. Mech. Phys. Sol. 48, 1701–1733 (2000)
5. Axelsson, O.: Iterative solution methods. Cambridge University Press (1994)
6. Bessis, M.: Corpuscules. Springer, Heidelberg (1974)
7. Bergman, D.J.: The Dielectric Constant of a Composite Material – A Problem in Classical Physics. Phys. Rep. C 43, 377 (1978)
8. Bianco, M., Bilardi, G., Pesavento, F., Pucci, G., Schrefler, B.A.: A frontal solver tuned for fully coupled non-linear hygro-thermo-mechanical problems. Int. J. Numer. Meth. Eng. 57, 1801–1818 (2003)
9. Brown, S., Abou-Chedid, G.: Yield behavior of metal powder assemblages. J. Mech. Phys. Solid. 42, 383–398 (1994)
10. Chandrasekharaiah, D.S., Debnath, L.: Continuum mechanics. Academic Press (1994)
11. Crank, J.: The mathematics of diffusion, 2nd edn. Oxford Science Publications (1975)
12. Cule, D., Torquato, S.: Electric-Field Distribution in Composite Media. Phys. Rev. B 58, R11829 (1998)
13. Davis, L.: Handbook of Genetic Algorithms. Thompson Computer Press (1991)
14. Demkowicz, L.: Computing with Hp-Adaptive Finite Elements. One- and Two-dimensional Elliptic and Maxwell Problems, vol. I. CRC Press (2006)
15. Demkowicz, L., Kurtz, J., Pardo, D., Paszynski, M., Rachowicz, W., Zdunek, A.: Computing with Hp-Adaptive Finite Elements. Vol. II. Frontiers: Three Dimensional Elliptic and Maxwell Problems with Applications. Chapman and Hall (2007)
16. Doltsinis, I.: Coupled field problems–solution techniques for sequential and parallel processing. In: Papadrakakis, M. (ed.) Solving Large–Scale Problems in Mechanics (1993)

17. Doltsinis, I.: Solution of coupled systems by distinct operators. Engrg. Comput. 14, 829–868 (1997)
18. Domas, F.: Eigenschaftprofile und Anwendungsübersicht von EPE und EPP. Technical report of the BASF Company (1997)
19. Donev, A., Cisse, I., Sachs, D., Variano, E.A., Stillinger, F., Connelly, R., Torquato, S., Chaikin, P.: Improving the density of jammed disordered packings using ellipsoids. Science 303, 990–993 (2004)
20. Donev, A., Torquato, S., Stillinger, F.: Neighbor list collision-driven molecular dynamics simulation for nonspherical hard particles-I. Algorithmic details. J. Comput. Phys. 202, 737 (2005a)
21. Donev, A., Torquato, S., Stillinger, F.: Neighbor list collision-driven molecular dynamics simulation for nonspherical hard particles-II. Application to ellipses & ellipsoids. J. Comput. Phys. 202, 765 (2005b)
22. Eber, S., Lux, S.E.: Hereditary spherocytosis–defects in proteins that connect the membrane skeleton to the lipid bilayer. Semin. Hematol. 41, 118–141 (2004)
23. Evans, E.A., Fung, Y.C.: Improved measurements of the erythrocyte geometry. Microvascular Research 4, 335 (1972)
24. Farhat, C., Lesoinne, M., Maman, N.: Mixed Explicit/Implicit Time Integration of Coupled Aeroelastic Problems: Three-Field Formulation, Geometric Conservation and Distributed Solution. Int. J. Num. Method. Fluid 21, 807–835 (1995)
25. Farhat, C., Lesoinne, M.: Two Efficient Staggered Procedures for the Serial and Parallel Solution of Three-Dimensional Nonlinear Transient Aeroelastic Problems. Comp. Meth. App. Mech. and Engrg. 182, 499–516 (2000)
26. Farhat, C., van der Zee, G., Geuzaine, P.: Provably second-order time-accurate loosely-coupled solution algorithms for transient nonlinear computational aeroelasticity. Comp. Meth. Appl. Mech. Engng. 195, 1973–2001 (2006)
27. Fish, J., Chen, W.: Modeling and Simulation of Piezocomposites. Comp. Meth. Appl. Mech. Engng. 192, 3211–3232 (2003)
28. Fleck, N.A.: On the cold compaction of powders. J. Mech. Phys. Solid 43, 1409–1431 (1995)
29. Forget, B.G., Cohen, A.R.: Thalassemia syndromes. In: Hoffman, B.E. (ed.) Hematology: Basic Principles and Practice, 4th edn., pp. 557–558. Elsevier, Philadelphia (2005)
30. Flynn, C.P.: Point defects and diffusion. Clarendon Press, Oxford (1972)
31. Gallagher, P.G.: Hereditary elliptocytosis: spectrin and protein 4.1R. Semin Hematol 41, 142–164 (2004a)
32. Gallagher, P.G.: Update on the clinical spectrum and genetics of red blood cell membrane disorders. Curr. Hematol. Rep. 3, 85–91 (2004b)
33. Gill, P., Murray, W., Wright, M.: Practical optimization. Academic Press (1995)
34. Goldberg, D.E.: Genetic algorithms in search, optimization and machine learning. Addison-Wesley (1989)
35. Goldberg, D.E., Deb, K.: Special issue on Genetic Algorithms. Comp. Meth. Appl. Mech. Engng. 186(2-4), 121–124 (2000)
36. Grant, et al.: A critical study of the open-ended coaxial line sensor technique for RF and microwave complex. J. Phys. E: Sci. Instrum. (1989)
37. Gu, C., Kim, M., Anand, L.: Constitutive equations for metal powders: application to powder forming processes. Int. J. Plast. 17, 147–209 (2001)

38. Gustafson, J.L.: Reevaluating Amdahl's law. Comm. ACM 31(5), 532–533 (1988)
39. Hashin, Z., Shtrikman, S.: A variational approach to the theory of effective magnetic permeability of multiphase materials. J. App. Phys. 33(10), 3125–3131 (1962)
40. Hashin, Z., Shtrikman, S.: On some variational principles in anisotropic and nonhomogeneous elasticity. J. Mech. Phys. Solid 10, 335–342 (1962)
41. Hashin, Z., Shtrikman, S.: A variational approach to the theory of the elastic behaviour of multiphase materials. J. Mech. Phys. Solid. 11, 127–140 (1963)
42. Hashin, Z.: Analysis of composite materials: a survey. ASME J. App. Mech. 50, 481–505 (1983)
43. Hill, R.: The Elastic Behaviour of a Crystalline Aggregate. Proc. Phys. Soc. Lond A65, 349–354 (1952)
44. Hector, L.G., Schultz, H.L.: The Dielectric Constant of Air at Radiofrequencies 7, 133–136 (1936)
45. Holland, J.H.: Adaptation in natural and artificial systems. University of Michigan Press, Ann Arbor (1975)
46. Jackson, J.D.: Classical Electrodynamics, 3rd edn. Wiley (1998)
47. Jikov, V.V., Kozlov, S.M., Olenik, O.A.: Homogenization of differential operators and integral functionals. Springer, Heidelberg (1994)
48. Kachanov, L.M.: Introduction to continuum damage mechanics. Martinus Nijoff, Dordricht (1986)
49. Kansaal, A., Torquato, S., Stillinger, F.: Diversity of order & densities in jammed hard-particle packings. Phys. Rev. E 66, 041109 (2002)
50. Kennedy, J., Eberhart, R.: Swarm Intelligence. Morgan Kaufmann Publishers (2001)
51. Kreher, W., Pompe, W.: Internal Stresses in Heterogeneous Solids. Akademie-Verlag, Berlin (1989)
52. Kröner, E.: Statistical Continuum Mechanics. CISM Lecture Notes, vol. 92. Springer, Heidelberg (1972)
53. Kunz, K.S., Luebbers, R.J.: The Finite Difference Time Domain Method for Electromagnetics. CRC Press (1993)
54. Lagaros, N., Papadrakakis, M., Kokossalakis, G.: Structural optimization using evolutionary algorithms. Comp. Struct. 80, 571–589 (2002)
55. Lesoinne, M., Farhat, C.: Free Staggered Algorithm for Nonlinear Transient Aeroelastic Problems. AIAA J. 36(9), 1754–1756 (1998)
56. Le Tallec, P., Mouro, J.: Fluid structure interaction with large structural displacements. Comp. Method. App. Mech. Engrg. 190(24-25), 3039–3067 (2001)
57. Lewis, R.W., Schrefler, B.A., Simoni, L.: Coupling versus uncoupling in soil consolidation. Int. J. Num. Anal. Metho. Geomech. 15, 533–548 (1992)
58. Lewis, R.W., Schrefler, B.A.: The finite element method in the static and dynamic deformation and consolidation of porous media, 2nd edn. Wiley Press (1998)
59. Lipton, R.: Optimal bounds on electric-field fluctuations for random composites. J. App. Phys. 88, 4287–4293 (2000)
60. Lipton, R.: Assessment of the local stress state through macroscopic variables. Phil. Trans. Math. Phy. Engrg. Sci. 361, 921–946 (2003)
61. Lipton, R.: Optimal lower bounds on the electric-field concentration in composite media. J. App. Phys. 96, 2821–2827 (2004)

62. Lipton, R.: Optimal lower bounds on the dilatational strain inside random two-phase elastic composites subjected to hydrostatic loading. Mech. Mat. 38, 833–839 (2006)
63. Luenberger, D.: Introduction to Linear and Nonlinear Programming. Addison-Wesley, Menlo Park (1974)
64. Malvern, L.: Introduction to the mechanics of a continuous medium. Prentice-Hall (1968)
65. Markenscoff, X.: Diffusion induced instability. Q. App. Mech. LIX (1), 147–151 (2001)
66. Markenscoff, X.: Instabilities of a thermo-mechano-chemical system. Q. App. Mech. LIX(3), 471–477 (2001)
67. Markenscoff, X.: On conditions of "negative creep" in amorphous solids. Mech. Mater. 35(3-6), 553–557 (2003)
68. Maxwell, J.C.: On the dynamical theory of gases. Philos. Trans. Soc. London 157, 49 (1867)
69. Maxwell, J.C.: A treatise on electricity and magnetism, 3rd edn. Clarendon Press, Oxford (1873)
70. Michaud, V.: Liquid state processing. In: Suresh, S., Mortensen, A., Needleman, A. (eds.) Fundamentals of Metal Matrix Composites (1992)
71. Michopoulos, G., Farhat, C., Fish, J.: Survey on Modeling and Simulation of Multiphysics Systems. J. Comp. Inf. Sci. Engrg. 5(3), 198–213 (2005)
72. Mura, T.: Micromechanics of defects in solids, 2nd edn. Kluwer Academic Publishers (1993)
73. Nye, J.F.: Physical properties of crystals. Oxford University Press (1957)
74. Nemat-Nasser, S., Hori, M.: Micromechanics: overall properties of heterogeneous solids, 2nd edn. Elsevier, Amsterdam (1999)
75. Oden, J.T., Zohdi, T.I.: Analysis and adaptive modeling of highly heterogeneous elastic structures. Comp. Meth. App. Mech. Engrg. 148, 367–391 (1997)
76. Onwubiko, C.: Introduction to engineering design optimization. Prentice-Hall (2000)
77. Park, K.C., Felippa, C.A.: Partitioned analysis of coupled systems. In: Belytschko, T., Hughes, T.J.R. (eds.) Computational Methods for Transient Analysis (1983)
78. Piperno, S.: Explicit/implicit fluid/structure staggered procedures with a structural predictor and fluid subcycling for 2D inviscid aeroelastic simulations. Int. J. Num. Meth. Fluid. 25, 1207–1226 (1997)
79. Papadrakakis, M.: Solving Large-Scale Problems in Mechanics. John Wiley and Sons (1993)
80. Papadrakakis, M.: Parallel Solution Methods in Computational Mechanics. John Wiley and Sons (1997)
81. Papadrakakis, M., Lagaros, N.D., Fragakis, Y.: Parallel computational strategies for structural optimisation. Int. J. Num. Meth. Engrg. (2003)
82. Papadrakakis, M., Lagaros, N., Thierauf, G., Cai, J.: Advanced solution methods in structural optimisation using evolution strategies. Engrg. Comput. J. 15(1), 12–34 (1998a)
83. Papadrakakis, M., Lagaros, N., Tsompanakis, Y.: Structural optimization using evolution strategies and neutral networks. Comp. Method. App. Mech. Engrg. 156(1), 309–335 (1998b)
84. Papadrakakis, M., Lagaros, N., Tsompanakis, Y.: Optimization of large-scale 3D trusses using Evolution Strategies and Neural Networks. Int. J. Space Struc. 14(3), 211–223 (1999a)

85. Papadrakakis, M., Tsompanakis, J., Lagaros, N.: Structural shape optimisation using evolution strategies. Eng. Optimization 31, 515–540 (1999b)
86. Papadrakakis, M., Lagaros, N., Tsompanakis, Y., Plevris, V.: Large scale structural optimization: Computational methods and optimization algorithms. Arch. Comput. Meth. Engrg., SOTA Rev. 8(3), 239–301 (2001)
87. Piperno, S., Farhat, C., Larrouturou, B.: Partitioned Procedures for the Transient Solution of Coupled Aeroelastic Problems – Part I: Model Problem, Theory, and Two-Dimensional Application. Comp. Method. App. Mech. Engrg. 124(1-2), 79–112 (1995)
88. Piperno, S., Farhat, C.: Partitioned Procedures for the Transient Solution of Coupled Aeroelastic Problems – Part II: Energy Transfer Analysis and Three-Dimensional Applications. Comp. Method. App. Mech. Engrg. 190, 3147–3170 (2001)
89. Rachowicz, W., Zdunek, A.: Automated multi-level substructuring (AMLS) for electromagnetics. Comp. Method. App. Mech. Engrg. 198(13/14), 1224–1234 (2009)
90. Rayleigh, J.W.: On the influence of obstacles arranged in rectangular order upon properties of a medium. Phil. Mag. 32, 481–491 (1892)
91. Reuss, A.: Berechnung der Fliessgrenze von Mischkristallen auf Grund der Plastizitätsbedingung für Einkristalle. Z. Angew. Math. Mech. 9, 49–58 (1929)
92. Schrefler, B.A.: A partitioned solution procedure for geothermal reservoir analysis. Comm. Appl. Num. Meth. 1, 53–56 (1985)
93. Schwan, C.: Linear and nonlinear electrode polarization and biological materials. Annals of Biomedical Engineering (1992)
94. Schwan, C.: Electrical properties of blood and its constituents: alternating current spectroscopy. Blut. 46(4), 185–197 (1983)
95. Sevostianov, I., Gorbatikh, L., Kachanov, M.: Recovery of information of porous / microcracked materials from the effective elastic / conductive properties. Mat. Sci. Engrg. A318, 1–14 (2001)
96. Stavroulakis, G.M., Papadrakakis, M.: Advances on the domain decomposition solution of large scale porous media problems. Comp. Method. App. Mech. Engrg. 198(21-26), 1935–1945 (2009)
97. Steinberg, M.H., Benz Jr., E.J., Adewoye, H.A., Hoffman, B.E.: Pathobiology of the human erythrocyte and it's hemoglobins. In: Hoffman, B.E. (ed.) Hematology: Basic Principles and Practice, 4th edn., pp. 442–452. Elsevier, Philadelphia (2005)
98. Taflove, A., Hagness, S.: Computational Electrodynamics: The Finite-Difference Time-Domain Method, 3rd edn. Artech Press (2005)
99. Tatzel, H.: Grundlagen der Verarbeitungstechnik von EPP-Bewährte und neue Verfahren. Technical report of the BASF Company (1996)
100. Torquato, S.: Random heterogeneous media: microstructure and improved bounds on effective properties. Appl. Mech. Rev. 44, 37–76 (1991)
101. Torquato, S., Lado, F.: Improved bounds on the effective elastic moduli of random arrays of cylinders. J. Appl. Mech. 59, 1–6 (1992)
102. Torquato, S.: Effective stiffness tensor of composite media I. Exact series expansions. J. Mech. Phys. Solid 45, 1421–1448 (1997)
103. Torquato, S.: Effective stiffness tensor of composite media I. Applications to isotropic dispersions. J. Mech. Phys. Solid 46, 1411–1440 (1998)
104. Torquato, S.: Random Heterogeneous Materials: Microstructure and Macroscopic Properties. Springer, New York (2001)

105. Tursa, E., Schrefler, B.A.: On consistency, stability and convergence of staggered solution procedures. Rend. Mat. Acc. Lincei, Rome S. 9. 5, 265–271 (1994)
106. Voigt, W.: Über die Beziehung zwischen den beiden Elastizitätskonstanten isotroper Körper. Wied. Ann. 38, 573–587 (1889)
107. Wang, X., Schrefler, B.A.: A multifrontal parallel algorithm for coupled thermo-hydro-mechanical analysis of deforming porous media. Int. J. Numer. Meth. Eng. 43, 1069–1083 (1998)
108. Widom, B.: Random sequential addition of hard spheres to a volume. J. Chem. Phys. 44, 3888–3894 (1966)
109. Wiener, O.: Zur Theorie der Refraktionskonstanten. Berichte über die Verhandlungen der Königlich-Sächsischen Gesellschaft der Wissenshaften zu Leipzig. Math.-Phys. Klasses, Band 62, 256–277 (1910)
110. Yee, K.: Numerical solution of initial boundary value problems involving Maxwell's equations in isotropic media. IEEE Trans. on Antennas and Propagation 14, 302 (1966)
111. Young. D. M.: Iterative methods for solving partial difference equations of elliptic type. Doctoral thesis. Harvard University (1950)
112. Zienkiewicz, O.C.: Coupled problems & their numerical solution. In: Lewis, R.W., Bettes, P., Hinton, E. (eds.) Numerical Methods in Coupled Systems, pp. 35–38. Wiley, Chichester (1984)
113. Zienkiewicz, O.C., Paul, D.K., Chan, A.H.C.: Unconditionally stable staggered solution procedure for soil-pore fluid interaction problems. Int. J. Num. Meth. Engrg. 26, 1039–1055 (1988)
114. Zohdi, T.I., Oden, J.T., Rodin, G.J.: Hierarchical modeling of heterogeneous bodies. Comp. Meth. App. Mech. Engrg. 138, 273–298 (1996)
115. Zohdi, T.I., Wriggers, P.: A domain decomposition method for bodies with microstructure based upon material regularization. Int. J. Solid. Struct. 36(17), 2507–2526 (1999)
116. Zohdi, T.I.: Simulation of time-discontinuous chemically-aided intergranular fracture. Comput. Mat. Sci. 24(4), 490–500 (2002)
117. Zohdi, T.I., Hutter, K., Wriggers, P.: A technique to describe the macroscopic pressure dependence of diffusive properties of solid materials containing heterogeneities. Comput. Mat. Sci. 15, 69–88 (1999)
118. Zohdi, T.I.: Some remarks on hydrogen trapping. Int. J. Fract. 106(2), L9–L14 (2000)
119. Zohdi, T.I., Wriggers, P.: A domain decomposition method for bodies with microstructure based upon material regularization. Int. J. Solid Struc. 36(17), 2507–2526 (1999)
120. Zohdi, T.I., Wriggers, P.: Aspects of the computational testing of the mechanical properties of microheterogeneous material samples. Int. J. Num. Meth. Engrg. 50, 2573–2599 (2001)
121. Zohdi, T.I., Wriggers, P.: Computational micro-macro material testing. Arch. Comput. Method. Engrg. 8(2), 131–228 (2001)
122. Zohdi, T.I., Wriggers, P., Huet, C.: A method of substructuring large-scale computational micromechanical problems. Comp. Method. App. Mech. Engrg. 190(43/44), 5639–5656 (2001)
123. Zohdi, T.I.: Computational optimization of vortex manufacturing of advanced materials. Comp. Method. App. Mech. Engrg. 190(46/47), 6231–6256 (2001)

124. Zohdi, T.I.: Genetic design of solids possessing a random-particulate microstructure. PTRS: Math. Phys. Engrg. Sci. 361(1806), 1021–1043 (2003)
125. Zohdi, T.I.: On the compaction of cohesive hyperelastic granules at finite strains. Proc. Royal Soc. 454(2034), 1395–1401 (2003)
126. Zohdi, T.I.: Encyclopedia chapter. In: Stein, E., de Borst, R., Hughes, T. (eds.) Homogenization Methods and Multiscale Modeling: Linear Problems. Encyclopedia of Computational Mechanics. John Wiley (2004)
127. Zohdi, T.I., Wriggers, P.: Introduction to computational micromechanics. Springer, Heidelberg (2005)
128. Zohdi, T.I., Wriggers, P.: Introduction to computational micromechanics, second Reprinting. Springer, Heidelberg (2008)
129. Zohdi, T.I.: An adaptive-recursive staggering strategy for simulating multifield coupled processes in microheterogeneous solids. Int. J. Num. Method. Engrg. 53, 1511–1532 (2002)
130. Zohdi, T.I.: Computational design of swarms. Int. J. Num. Method. Engrg. 57, 2205–2219 (2003)
131. Zohdi, T.I.: Modeling and simulation of a class of coupled thermo-chemomechanical processes in multiphase solids. Comp. Method. App. Mech. Engrg. 193(6/8), 679–699 (2004)
132. Zohdi, T.I.: Modeling and direct simulation of near-field granular flows. Int. J. Solid. Struc. 42(2), 539–564 (2004)
133. Zohdi, T.I.: Computation of the coupled thermo-optical scattering properties of random particulate systems. Comp. Method. App. Mech. Engrg. 195, 5813–5830 (2006)
134. Zohdi, T.I.: Computation of strongly coupled multifield interaction in particle-fluid systems. Comp. Method. App. Mech. Engrg. 196, 3927–3950 (2007)
135. Zohdi, T.I., Kuypers, F.A.: Modeling and rapid simulation of multiple red blood cell light scattering. Proc. Roy. Soc. Inter. 3(11), 823–831 (2006)
136. Zohdi, T.I., Kuypers, F.A., Lee, W.C.: Estimation of Red Blood Cell volume fraction from overall permittivity measurement. Int. J. Engrg. Sci. Online 48, 1681–1691 (2010)
137. Zohdi, T.I.: On the computation of the coupled thermo-electromagnetic response of continua with particulate microstructure. Int. J. Num. Method. Engrg. 76, 1250–1279 (2008)
138. Zohdi, T.I.: Joule-heating field phase-amplification in particulate-doped dielectrics. Int. J. Engrg. Sci. 49, 30–40 (2011)
139. Zohdi, T.I.: Simulation of coupled microscale multiphysical-fields in particulate-doped dielectrics with staggered adaptive FDTD. Comp. Method. App. Mech. Engrg. 199, 79–101 (2010)
140. Zohdi, T.I.: Estimation of electrical Load-shares For sintering of powder mixtures. Proceedigs of the Royal Society (in press)

Index

The manufacturer's authorised representative in the EU is Springer Nature Customer Service Centre GmbH, Europaplatz 3, 69115 Heidelberg, Germany. If you have any concerns regarding our products, please contact ProductSafety@springernature.com

Printed and bound by CPI Group (UK) Ltd, Croydon, CR0 4YY
30/09/2026
02225602-0001